THE IMPORTANCE OF *Being Observant*

A COLLECTION OF CASE STUDIES TO TEACH THE FUNDAMENTALS OF BIOLOGY

JON MILHON • SCOTT KINNES • JOSHUA MORRIS • JÜRGEN ZIESMANN

Printed in the United States of America.

ISBN: 1-60250-056-8
978-1-60250-056-3

59 Damonte Ranch Parkway, Suite B284 • Reno, NV 89521 • (800) 970-1883

www.benttreepress.com

Address all correspondence and order information to the above address.

Table of Contents

Introduction

We came upon the idea of using case studies to teach the fundamental principles of biology because, all too often, our students claimed biology was uninteresting. Ironically, teaching methods designed to engage students, like the use of case studies, have been reserved for students who are self-motivated and possess some background biology knowledge. Unfortunately, it is the students with the least biology background who need to be engaged the most. The purpose of writing these cases is to show students how our everyday lives are regularly impacted by interesting and relevant biological concepts. Our hope is that the cases push students into the unknown, so they have to go searching for answers. The library and the Internet are full of reputable sources of information in every field of biology. Sure, there are always some quacks out there, but if you use care, you can find reliable information on almost any topic. In finding answers, we think you'll agree that biology is, indeed, very interesting and applicable.

About the Authors

Professional Bio: Jon Milhon

Jon Milhon began his academic career at Azusa Pacific University as an undergraduate student. Upon graduation, he pursued a Ph.D. at the University of Southern California, studying transcriptional activation by glucocorticoid receptors in Dr. Michael Stallcup's lab. He then went on to do postdoctoral work on the parasitic trematode *Schistosoma mansoni* in the lab of Dr. James Tracy at the University of Wisconsin, Madison. Dr. Milhon's career has come full circle, as he now teaches and performs research with undergraduate students at his alma mater, Azusa Pacific University. The idea of writing a set of case studies geared toward undergraduates was developed while trying to find ways to excite, engage, and challenge students in his Fundamentals of Biology course.

Professional Bio: Joshua Morris

Joshua Morris began his academic career at California State Polytechnic Institute, Pomona, where he earned a B.S. degree in biology. Upon completion, he pursued a Ph.D. in Molecular, Cell, and Developmental Biology at the University of California, Los Angeles. Working in the lab of Dr. Volker Hartenstein, he studied the neural development of a platyhelminth worm, *Macrostomum lignano.* Dr. Morris is now an Assistant Professor at Azusa Pacific University, where he teaches genetics and is continuing research on *Macrostomum* with undergraduate students.

Professional Bio: Jürgen Ziesmann

Jürgen Ziesmann studied biology at Bayreuth University, Germany, where he graduated and then completed his Ph.D. in the field of sensory physiology of animals. He then held several postdoctoral positions at Rothamsted Experimental Station, UK, the Max-Planck Institute for Behavioral Physiology at Seewiesen, Germany, and the University of Maryland at Baltimore City, USA. As a postdoctoral fellow, he researched the olfactory sense of animals, including nematodes, insects, and mice. Since 2002, Dr. Ziesmann has taught Anatomy and Physiology at Azusa Pacific University and continues research in the field of olfaction.

Professional Bio: Scott S. Kinnes

Scott S. Kinnes did his undergraduate work at Belhaven College in Jackson, Mississippi, where he majored in Biology. He continued his education at Duke University in Durham, NC, where he earned a Masters in forestry and a Doctorate in soil microbiology. He has been teaching at the university level for 25 years and is currently a Professor at Azusa Pacific University. There he teaches in the areas of microbiology, ecology, and ethics of science. He has published several laboratory manuals in the area of general biology for both biology majors and non-majors and has taught these courses at a number of institutions.

The Importance of Being Observant

A Case Study of Penicillin and the Scientific Method

Alex had a problem, the same problem that microbiology students have every semester as they try to get **bacteria** to grow on agar in Petri dishes. All too frequently, dust and **fungal spores** fall out of the air and into the dish, contaminating the bacterial cultures. When this happens to microbiology students, they simply moan and groan, get a lecture from the professor about being more careful next time, and throw the plate away. When professional microbiologists do this, they simply fuss and cuss, throw the plates away, and repeat the experiment.

Alex's problem was that, all too often, his plates of various species of ***Staphylococcus*** **bacteria** also became contaminated with fungal spores. Now, in the 1920s, his studies on *Staphylococcus* were very important, as this group of bacteria causes some pretty significant diseases in humans. Many types of skin infections, wound infections, food poisonings and other types of infections are due to this **genus**. We now know that even **Toxic Shock Syndrome** is due to ***S. aureus***, one of the most pathogenic species in the genera. Therefore, his anger at continually having his *Staphylococcus* plates ruined would be understandable, as it was ruining his important work.

Fortunately for all of us, Alexander Fleming shared a trait with many important scientists—he was very **observant**. So, while you and I may have just tossed away the contaminated plates and started all over again, he started to notice that occasionally wherever the fungal contamination was, there was no bacterial growth. Rather than throw the plates away, Fleming started to experiment with this particular type of contamination in an effort to determine what was going on. He thus switched from the question he was originally working on to a new **question**: Why were the bacteria not growing around the fungus? Based upon his previous work in microbiology, his guess as to what was happening was that the fungus was making some chemical that was preventing the bacteria from growing around it. We call such an educated guess a **hypothesis**.

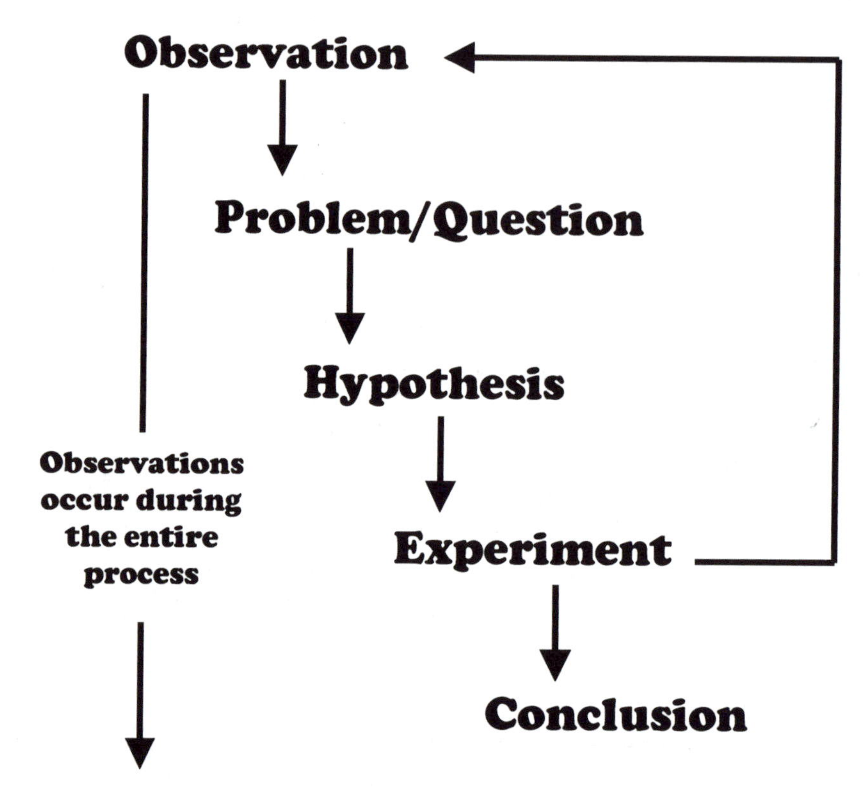

In order to prove this idea, Fleming had to develop an **experiment** that would test his hypothesis. Eventually he was able to identify that such a chemical was being produced, isolate an extract of the chemical, and name it **penicillin** after the genus of fungus, ***Penicillium,*** that was making it. Thus, he eventually came to a **conclusion** that set the stage for the development of **antibiotics**, which continue to save millions of lives each year. How different our lives might be if Sir Alexander Fleming had not been so good at using the **scientific method**!

This method allows scientists, and anyone else trying to solve problems, to logically proceed through a number of steps to solve problems in a way that can be repeated by others. It is the basic set of steps that allows science to proceed in solving problems and answering questions that arise. This same sequence of steps can be used to solve any sort of problem, such as the one I had to solve with my old, somewhat faithful, car.

For some reason, every time it rained, my poor old car would not start. Living in Southern California, this was not a big problem. However, at certain times of the year, it really does rain in Southern California and it got to be quite a pain, as I had to let the car sit, wherever it happened to die, until nice weather came and it would start right up. So I had a question based upon my observations: Why would my car not start in the rain?

Now, not being very knowledgeable about cars, I could not even make an educated guess on my own, so I had to start doing some reading on the subject, ask a few friends who I knew worked on their own cars, and bother the guy down at the auto parts store. Based upon this "research" I came to a "hypothesis": there was a crack in my distributor cap. The distributor cap covers the electrical connections that feed the spark plugs. My hypothesis was that when water gets under the cap, the electrical connections short out. My "experiment" consisted of buying a new cap, installing it, and running the car through a local car wash. Unfortunately, it turned out that my hypothesis was incorrect and the experiment proved that because the car would not start after the car wash!

So…back to the drawing board, or to the local mechanic, in this case. Soon, I came up with a new hypothesis: I had a bad rotor in the distributor (Don't worry about what it does; it is kind of complicated). This generated a new experiment, which involved purchasing and installing the new rotor in the new cap and hitting the car wash again. Result? A stalled car!

Finally, more research led me to the hypothesis that the ignition wires, running from the distributor to the spark plugs (they carry the electrical current), might be old and cracked. This would mean that they would work fine in dry weather but that any kind of moisture would cause the wires to malfunction and the car to not start. Experiment? Buying and installing new ignition wires. Result? It started and kept running no matter what the weather was! Conclusion? Changing the ignition wires will, in this case, prevent problems in wet weather. Eureka!

Now, while this demonstrates the steps in the scientific method, it also demonstrates one of the outcomes that may occur. If your first hypothesis does not lead to an answer to the question, you may have to develop a new hypothesis, based upon more research, and perform a new experiment to test that hypothesis. The story goes that Thomas Edison had to do this several thousand times before he found the right material to create the first light bulb.

Alexander Fleming's story illustrates another important aspect of the scientific method. This is the idea that some great discoveries are made because the scientist was not so intent on answering his or her first problem and was willing to let the experimental research lead him or her in new, more interesting and ultimately, more important directions. This means that, for the time being, Fleming allowed his new observations to create a new problem, which led to a new hypothesis, new experiments, and new conclusions.

Finally, while Fleming is credited with the discovery of the first antibiotic, the story is more complex than that, and further illustrates another important aspect of how science operates. While Fleming "accidentally" discovered and extracted penicillin from the *Penicillium* mold, there was a long history of previous observations made by other scientists that set the stage for his discovery, dating back decades earlier. Joseph Lister noted that if a urine sample became contaminated with mold, a type of fungus, no bacteria would grow in the sample in 1871. In 1874, William Roberts noted that *Penicillium* cultures rarely became contaminated with bacteria. Louis Pasteur, known for numerous discoveries in the field of microbiology, also noticed that one dangerous bacterium he was working with, that caused the disease Anthrax, would not grow in the presence of certain fungi. Now, they never were able to identify the fungi causing this, but it was one more link in the chain leading to Fleming's work. Finally, a student working on his doctoral dissertation in 1897, Ernest Duchesne, actually observed that the common bacteria ***Escherichia coli*** found in our **intestines**, was killed if it was grown with the fungus *Penicillium glaucum*. Other scientists also contributed to the background data that Fleming would have known when doing his work in the 1920s. Their work, therefore, contributed to Fleming's accomplishment.

This process of building upon the work of others is a very important one in science and was actually responsible for the eventual creation of a functional antibiotic. While Fleming set the stage for this in 1928 when he was able to isolate the chemical he called penicillin, he was never able to create a pure, concentrated extract of the chemical. This would not be done until a team of scientists, working some ten years later, was able to accomplish this feat. This team, led by Howard Florey and including Ernest Chain and Norman Heatley, was finally able to collect and purify sufficient quantities of the chemical to test in laboratory animals exposed to **infective pathogens**. When these animals did not come down with the specific disease, it scientifically demonstrated the effectiveness of penicillin and led, many years later, to its commercial production and widespread use...just in time for WWII. Without penicillin, infected wounds during WWII would have taken millions of lives.

Were it not for the fact that these scientists used the scientific method, recorded their data, and reported their findings, none of this would have happened as early as it did. The scientific method was thus responsible, ultimately, for a discovery that has saved millions of lives over the years...and can just as easily help you figure out a good solution to any problem you may be facing!

Questions about this case:

1. What is Toxic Shock Syndrome?
2. When designing an experiment, what does it mean to have a control? In the experiment by Florey, Chain, and Heatley mentioned above, how might they have controlled their experiment?
3. If a hypothesis is an "educated guess" what is a theory? Can there be multiple theories that explain the observations?
4. If I had to use the scientific method when fixing my car, surely you have used the scientific method as well. How would you employ this thought process if you try to turn on your desk lamp and it doesn't light? What would be your first hypothesis? If the desk lamp example doesn't work for you, choose one of your own.

Questions to go deeper:

1. What are some basic ways bacteria and fungi differ?
2. Are there infective pathogens that are not bacterial? If so, what are they?
3. Didn't the case mention that we *normally* have bacteria in our intestines? How does this work without making us sick?

References:

1. Anonymous. 2003. The Discovery of Penicillin. Nobel Foundation
   Nobelprize.org/medicine/educational/penicillin/readmore.html
2. Slowiczek, F. and P. Peters. 2006. Discovery, Chance and the Scientific Method. Access Excellence. www.accesssexcellence.org/AE/AEC/CC/chance.html
3. Wong, G. 2003. Penicillin. Univ. of Hawaii
   www.botany.hawaii.edu/faculty/wong/BOT135/Lect21b.htm

The Smell of Homemade Bread

A Case Study of Anaerobic Respiration

Watching my mother cook was always a learning experience. She knew a number of tricks that made cooking much faster and easier. I can vividly remember watching her bake bread. She always started the process by doing what she called "proofing" the yeast. She would add the yeast and some sugar to warm water. She would then let it sit for 5 minutes or so. As we were waiting, she explained that you need warm water because hot water would kill the yeast and cool water would not activate it. After the 5 minutes, she would look for bubbles to indicate that the yeast were alive. If the yeast did not make bubbles, then she would start over with a new batch of yeast because without living yeast, the bread would not rise.

It wasn't until later that this really made sense to me. Yeast are living, single-celled organisms. These cells are much like animal cells, in that they have a **nucleus** and similar **organelles**. Scientists call cells that have a nucleus and organelles **eukaryotes**. On the other hand, bacteria are called **prokaryotes** because they are single-celled organisms that do not have a nucleus or organelles. Since yeast are eukaryotes, they have similar chemical processes to animal cells. In general, cells prefer to use a sugar molecule containing six carbon atoms, called **glucose,** as a source of energy. In fact, most animals store long polymers of glucose called **glycogen** in cells for immediate use as energy. This glucose is broken down in a multi-step process called **glycolysis** in order to make the adenosine triphosphate (**ATP**) that the cell needs to function. ATP serves as the "energy currency" of the cell. When the cell needs energy to perform a task, the last of the three phosphates (remember adenosine **tri**phosphate) is removed, which releases the needed energy. When this one phosphate is removed (and energy is released) adenosine diphosphate **(ADP)** is produced. But, once the phosphate is removed, how do we convert ADP back to ATP? Well, glycolysis requires 2 molecules of ATP to get it started, but the process subsequently produces 4 ATP. Therefore, the process of glycolysis provides enough energy, from the breakdown of glucose, to convert 4 ADP back to 4 ATP. However, since glycolysis requires 2 ATP to start, the net gain of ATP is actually two. Glycolysis also produces 2 **activated carriers.** These molecules are "charged" by carrying the **electrons** that are generated during glycolysis. You can think of these carriers being charged in the same sense that rechargeable batteries are charged (and recharged). These carriers will help produce additional ATP using the electron transport chain, which will be discussed later. Basically, the process of glycolysis can be summed up with the following "equation": glycolysis breaks glucose down (with 6 carbon atoms) into 2 **pyruvate** molecules, each made of 3 carbon atoms, and leaves the cell with 2 net ATP and 2 charged activated carriers (see the actual equation below). However, this is where metabolism in yeast cells and animal cells diverge. Yeast are very simple: they have modest energy requirements. A net gain of 2 ATP molecules from breaking down glucose is sufficient for survival. Therefore, yeast convert the 2 pyruvate molecules into 2 molecules of **carbon dioxide** (CO_2) and 2 molecules of **ethanol.** You can see this process outlined in the figure below. Consequently, when my mother was looking for the bubbles, she was simply looking for the CO_2 that the yeast would create though glycolysis!

We call the breakdown of glucose **respiration** (which is not to be confused with the process of breathing). Yeast respiration would therefore be considered **anaerobic** because it does not require oxygen. As a result, anaerobic respiration could conceivably take place in an air-tight place like a capped bottle. Look again at the figure and notice that in the process of anaerobic respiration, both CO_2 and ethanol are produced. We call this **fermentation.** It should now be clear how fermented beverages

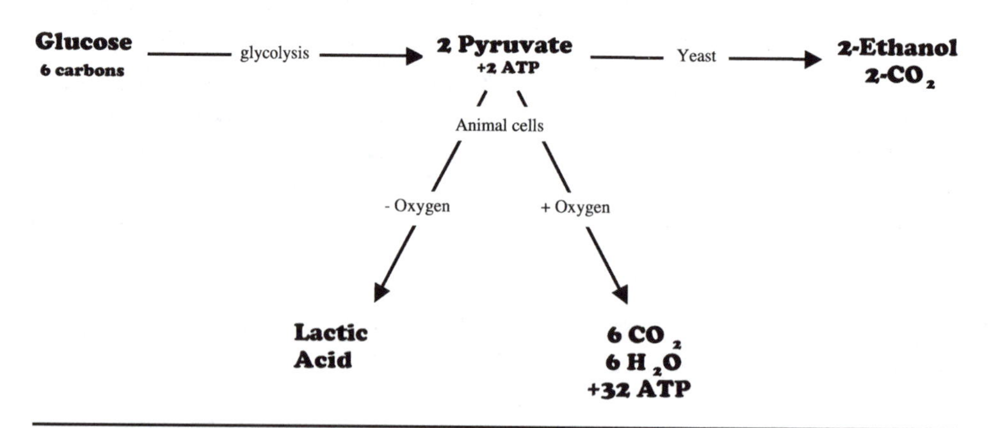

get both their alcohol and their natural carbonation, since yeast is used in their production.

As mentioned above, animal cells handle pyruvate very differently than yeast. Simply, gaining 2 net ATP by breaking down glucose into pyruvate (and subsequently into ethanol) does not provide enough energy for animal survival. Therefore, animal cells use **aerobic** respiration, which breaks the 2 pyruvate molecules into CO_2 and water. This further breakdown of pyruvate requires two processes called the **Kreb's Cycle** and the **electron transport chain** (again, see the figure). As the Kreb's Cycle breaks down the pyruvate molecules and releases CO_2, more activated carriers are charged. These charged carriers, plus the charged carriers from glycolysis, are then transported to the **mitochondrion.** Once there, they are processed though the electron transport chain. This produces a tremendous amount of ATP (32, to be exact, per glucose molecule). One may wonder what makes this respiration aerobic. It is the requirement for oxygen in the final step in the electron transport chain. Oxygen carries the electrons away from the electron transport chain. If oxygen is not present, glycolysis continues to produce 2 ATP per glucose molecule, but the Kreb's Cycle and the electron transport chain are shut down.

Animal cells do not survive very long under anaerobic conditions. You can see the result of putting human cells in anaerobic conditions by sprinting 400 meters. A lengthy sprint requires more oxygen in the cells of the leg muscles than can be delivered. If you have ever done this, then you know that once you are finished, your legs feel as heavy as lead and your muscles feel as if they are burning. These sensations are easily explainable. Without oxygen, only modest amounts of ATP are being produced in your leg muscles. Since sprinting requires so much energy, you are using ATP faster than it can be produced. This quickly leads to a shortage. Without ATP, muscles cannot contract efficiently, so your legs feel very heavy. In addition, the burning sensation is ultimately the result of the build-up of lactic acid. Without enough oxygen, the cells in your legs cannot process pyruvate through the Kreb's Cycle and the electron transport chain fast enough. The cells in your legs also cannot function anaerobically: they are unable to break down pyruvate into ethanol, as yeast can. Instead, they turn the excess pyruvate into **lactic acid**. Although lactic acid is not a very strong acid, it still results in a burning sensation in your legs when you perform a long sprint. Notice that during this sprint, the legs are primarily affected. Other muscles and organs are not working nearly as hard and are therefore not forced into anaerobic respiration, like the leg muscles are.

Fermentation has been used for many centuries in the manufacturing of food products. It is also currently the focal point for the development of alternative fuels. Automobiles can easily be adapted to run on ethanol, and as you have just seen, ethanol is easy to produce. To manufacture ethanol on a large scale, all you really need is some source of sugar (like sugarcane or corn) and yeast! It's cheap to produce and completely renewable. Imagine the economic and environmental impact of significantly reducing our oil usage to the point of not having to rely on imported oil.

I would have never thought that as I looked into a bowl of bubbling yeast as a kid, I was actually watching a process that answered so many everyday questions and could solve such significant problems.

Questions about this case:

1. Athletes who do long sprints require a significant rest after a race. What are the muscle cells in their legs doing with the lactic acid during this period of rest?

2. For years, gyms have offered "aerobics" classes. I can remember with some laughter the instructor asking if we could "feel the burn". What is the irony in this question?

3. If fermentation is going on during the process of bread-rising, why doesn't bread taste alcoholic?

4. Most people know that cyanide is very toxic. What you may not know is it has a major effect on the process of glucose breakdown. Why is cyanide so toxic?

Questions to go deeper:

1. How is aerobic respiration related to photosynthesis?

2. Is it realistic to think that we could produce enough ethanol by fermentation to drastically reduce our gasoline usage?

3. Under completely anaerobic conditions, the brain is the first organ to be significantly damaged. Why?

Be Careful How Much Electricity You Use

A Case Study of Photosynthesis

In March of 2004, the police were granted a search warrant to inspect a home in a quiet neighborhood in Carlsbad, California. The residents, a family of five including three children ages 13, 8, and 6, were suspected of growing marijuana for distribution. The key piece of evidence that resulted in the search warrant was a higher-than-average electrical use! As the police executed the warrant, it quickly became clear that the family was not growing marijuana at all, they just used 2-4 times more electricity than their neighbors did. At the home, the police were apologetic, but claimed that if they had to do it again, they would not have done anything differently. Why do the police use records of electrical usage as evidence for growing marijuana? The answer is quite simple. To grow outdoor plants inside, a special light bulb called a **"grow light"** is needed. Grow lights require significantly more electricity than normal bulbs. People who grow marijuana indoors use many of these bulbs and keep them on 24 hours a day.

The bigger question is why do outdoor plants being grown indoors need "grow lights" at all? Why don't regular fluorescent or incandescent bulbs work just as well for growing plants? The answer is found in understanding how plants use light and what type of light they use. Inside plant cells are organelles called **chloroplasts,** which contain **chlorophyll**. Chlorophyll is a protein that comes in two varieties, *a* and *b*. Together, the two types of chlorophyll absorb most visible light except yellow/green (see figure). Since these colors are not absorbed by plants, they are reflected, causing plants to be green in color. The artificial light emitted from a typical incandescent or fluorescent light bulb does not contain the full spectrum of visible light. Although artificial light looks white, it lacks an important part of the blue light that is absorbed by chlorophyll *b*. Without this blue light, the chlorophyll cannot fully do its job.

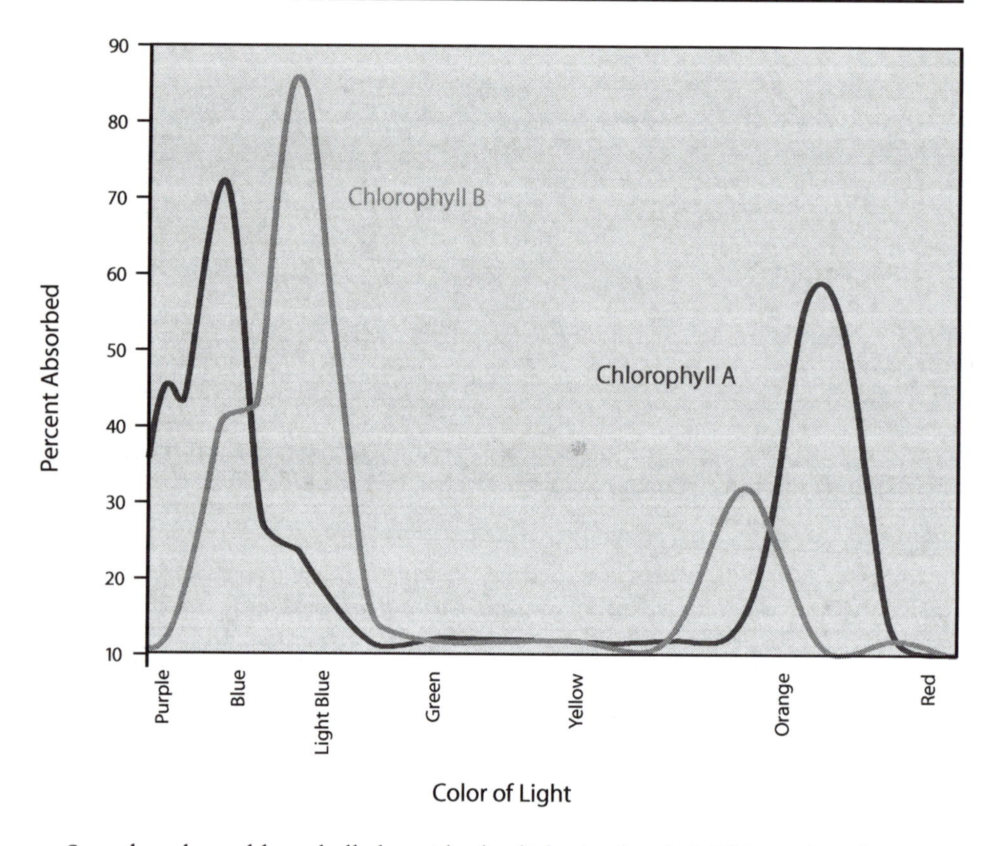

So, what does chlorophyll do with the light it absorbs? Ultimately, plants use chlorophyll to absorb sunlight in order to make carbohydrates for a fuel source. Plant cells use the light absorbed by chlorophyll to produce enough adenosine triphosphate **(ATP)** to produce carbohydrates, without needing to break down other complex molecules to do it. Animal cells can also use ATP to make carbohydrates, but to get that ATP, they must break down other complex molecules. This is why plants are called "**producers**" or "**autotrophs.**" You may be asking, how does ATP help in this process? ATP is a molecule that can release energy when one of its phosphates is removed. Basically, it is used as the energy currency; the energy that it releases is used to do some activity in the cell. To make carbohydrates from sunlight requires two distinct steps: one that requires sunlight and makes ATP, and a second that uses the ATP to supply the energy to make the carbohydrate. These two steps are called the **light dependent reaction** and **light independent reaction**, respectively. In the light dependent reaction, chlorophyll absorbs a **photon** of light, and in doing so, it breaks up a water molecule (releasing oxygen). This starts a set of chemical reactions that produce both ATP and **NADPH.** NADPH is what is called a **charged electron carrier**. The purpose of these charged carriers is to produce additional ATP. In the light independent reaction, the ATP that is produced by the light dependent reaction is then used along with six carbon dioxide (**CO_2**) molecules to produce one molecule of **glucose.** Glucose has a chemical formula of $C_6H_{12}O_6$. The reactions below describe the combination of the light dependent and the light independent reactions. Both reactions run in the same cell, and the result is the sum of adding the two reactions together (shown as the last equation). Notice that in the overall reaction, water, carbon dioxide, and light are needed to produce oxygen and glucose. The overall reaction also shows that the net ATP produced and consumed are equal. Therefore, ATP does not show up in the final reaction, since what is made in the first reaction is used in the second one.

As shown below, carbon dioxide is also needed for the photosynthesis to occur. However, since plant leaves are covered with a **waxy cuticle,** gases do not freely move into or out of leaves. How then does the plant take in CO_2? Well, it turns out that leaves also have small openings called **stomata,** which are designed to exchange gases (bring in CO_2 and release O_2), and can be opened and closed. Just as humans breathe in oxygen and breathe out carbon dioxide, plants bring in carbon dioxide and release oxygen through the stomata.

Light Dependent	**$12\ H_2O + \text{light} \rightarrow \text{ATP} + 6\ O_2$**
Light Independent	**$\text{ATP} + 6\ CO_2 \rightarrow C_6H_{12}O_6 + 6\ H_2O$**
Overall Reaction	**$6\ H_2O + 6\ CO_2 + \text{light} \rightarrow C_6H_{12}O_6 + 6\ O_2$**

We have also stated that plants are green because chlorophyll reflects green light. Molecules that reflect specific colors of light, like chlorophyll, are often called **pigments**. When **deciduous trees** begin to lose their leaves, they can go through brilliant color changes. In these situations, the chlorophyll is being broken down so it can no longer absorb and reflect light. Many trees have other pigments like **anthocyanin**, which appears to be red in color and serves as a protection from **UV light**. For the growing season, so much chlorophyll is made that it masks the presence of the other pigments. As autumn arrives, chlorophyll is broken down and the other pigments become solely responsible for absorbing and reflecting light. The colors that leaves turn in the fall depend on the other pigments the leaves make.

Not all plants require a grow light to be grown indoors. In fact, your local nursery should have a nice selection of "house plants" that can be grown indoors with no special treatment. Should you decide to grow an outdoor plant indoors, you should consider a grow light. Just be aware that switching all of your light bulbs to grow lights will be expensive, give off an odd bluish light, and will make the police suspicious.

Questions about this case:

1. Where do we find the plants that perform the highest percentage of the Earth's photosynthesis?
2. If plants use the ATP they make from sunlight to produce glucose, where do they get the ATP to do all of the other jobs around the cell that require energy?

Questions to go deeper:

1. If plants are called producers or autotrophs, what are humans called?
2. Hemoglobin in red blood cells is a pigment that reflects red light when oxygen is present. Why do you suppose bruises start out being red in color and then change to black, purple, blue, and even green?

Reference:

1. http://www.nctimes.com

Nate Doesn't Worry About His Hair Falling Out!

A Case Study of Cancer and Chemotherapy

I hadn't seen Nate for a couple of years, but when I heard the news of his cancer, I immediately made plans to visit him in the hospital. I was surprised by how good Nate looked for a guy who had just had his entire **stomach** and pieces of **esophagus** and **small intestine** removed. He explained that over the past two months, his appetite decreased to the point where he was nearly unable to eat and was in tremendous pain. Therefore, he was relieved to be eating even a small meal of soup and pudding without much pain. Nate had been diagnosed with stomach cancer after a lengthy battle to determine the source of his pain. Because the surgeon also found that this cancer had spread to his **pancreas** and **spleen**, it was determined that Nate would undergo **chemotherapy** after the surgery. Nate is not the type of guy who gets scared by much, and the thought of chemo did not seem to bother him. During the surgery, the surgeon fitted him with a plastic port so the chemo could be injected easily. I noticed that Nate was a victim of male pattern baldness, and the hair that he still had was cut very short. The prospect of losing his hair from the chemo gave us a good laugh.

We all seem to know the name "chemotherapy," but we rarely know how it works. The word really refers to any type of drug treatment for any therapeutic reason. However, it has been adopted by the cancer field and used almost exclusively in that context. Anytime a drug is given to a patient to kill something living in them, it is important that the drug's **mode of action** (how it kills "the problem") doesn't kill the normal cells, as well. This means that an effective chemotherapy for a cancer would kill the cancer cells, but not the normal human cells around it. The problem is that cancer cells are human cells. Fortunately, cancer cells do have characteristics that make them different from most human cells. Two commonly used anti-cancer drugs used today are **vinblastine** and **vincristine**. These chemicals were discovered in the periwinkle plant and have been shown to kill a variety of cancer cell types. In order to understand how the drugs vinbastine and vincristine work, an understanding of cell biology is essential.

Cells are self-contained units of life that can perform a variety of different tasks. The human body has numerous cell types, each with a specific function. The individual functions of a particular cell occur in specific compartments within the cell. These compartments are called **organelles**. The liquid part of the cells surrounding the organelles is called the **cytoplasm**. The cytoplasm is home to thousands of proteins, which are necessary to accomplish tasks, such as breaking down carbohydrates and the sending and receiving of molecular signals. Cells also have a "life cycle." Cells normally progress through the well-described cycle that involves four stages. Cells that have just divided are in a phase called **G1**. At this point, the cell will determine if conditions are favorable to continue in the cell cycle. If the cell chooses this path, then it will divide again. If conditions are not favorable, then the cell will stay in this stage until conditions improve. From the G1 phase, the cell will enter the **S-phase** (synthesis phase). This is where the cell will make identical copies of all 46 chromosomes. After leaving the S-phase, the cell will go through a phase called **G2,** where it will prepare for the process of cell division. The stage where the cell will physically divide is called **mitosis,** or **M-phase**. During this phase, each of the two new **daughter cells** will receive one copy of the 46 chromosomes that were duplicated in S-phase. They will also receive about half of the original organelles and cytoplasm. So what does this have to do with chemotherapy? Vinblastine and vincristine inhibit a key part of the cell cycle.

Within the cytoplasm of cells, there are three major types of filament that form the structural components of the cell. One of them, called **microtubules,** is special because they can be assembled and disassembled very rapidly. While we find this type of filament in a number of different cell types, microtubules are used universally when cells divide. This is because microtubules form the fibers (**spindle fibers**) that help to pull the chromosomes into the new daughter cells. Microtubules are made from subunits called **tubulin.** Vinbalstine and vincristine bind to tubulin and prevent microtubule formation, effectively halting mitosis. In this way, these two drugs are particularly effective in killing rapidly dividing cells, which is a characteristic of aggressive cancer cells.

If you have any friends or family who have suffered through chemo, you know there are significant side effects. We stated earlier that an effective chemotherapeutic drug needs to kill cancer cells, but not normal cells. To accomplish this, the drug would need to interfere with some characteristic that cancer cells have but normal human cells do not. Vinblastine and vincristine target rapidly dividing cells by disrupting microtubule formation. Rapid division is a characteristic that cancer cells possess; however, there are normal cells that are rapidly dividing, as well. These rapidly dividing normal cells need microtubules, as well. This is why there are many side effects to this type of chemotherapy. Nate is not concerned about his hair falling out, although it is very traumatic for most people. As you have probably guessed, cells in our hair follicles are rapidly dividing and die off in the presence of these drugs. Fortunately, after chemo is finished, these cells will grow back and produce more hair. Another side effect of vinblastine and vincristine treatments is intense nausea and vomiting. The cells that line your intestinal tract are also rapidly dividing and are killed by this type of chemotherapy. More significantly, these drugs can kill the actively dividing cells in the bone marrow that produces white blood cells. These drugs are also toxic to some neurons. Clearly, these are very toxic drugs, and they are administered at doses that can kill cancer cells. However, these doses will then cause near fatal side effects.

After Nate's surgery, he was allowed to go home and rest for about a month. This allowed him to gain some weight and strength so he can do battle with the very thing designed to save his life, chemotherapy.

Questions about this case:

1. Make a list of some of the most basic organelles and their functions.
2. Do you find it interesting that a chemotherapeutic drug was discovered in a common plant like the violet? How would you guess drugs like this were discovered? How are scientists trying to discover new drugs?
3. What are some of the other functions of microtubules?
4. If mitosis occurs in M-phase and DNA replication occurs in S-phase, what important events occur in G1 and G2?
5. We mentioned that when cells are not ready to progress through the cell cycle, they often stop in G1. What might happen if the cells lose the ability to stop?

Questions to go deeper:

1. What are the other two cellular filaments not specifically mentioned here? What are their general functions?
2. How would having no stomach affect Nate's digestion?
3. What would be the outcome of the side effect when bone marrow cells are killed? Why?

Reference:

1. http://users.rcn.com/jkimball.ma.ultranet/BiologyPages/G/GeneTherapy.html

This case is dedicated to my friend Nate Rawlings who passed away on August 26th, 2007.

Steve and His Glass Eye

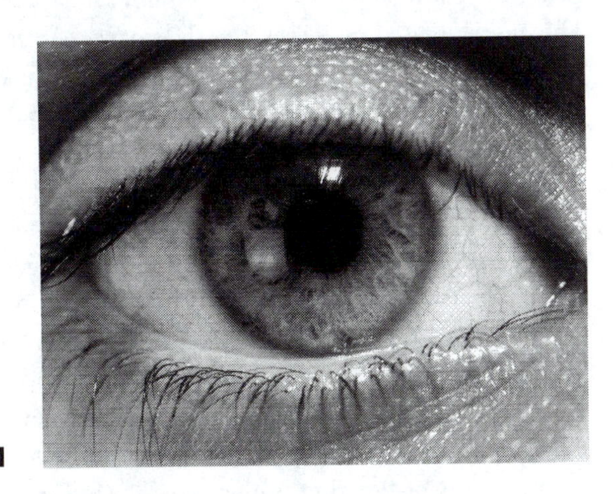

A Case Study of Retinoblastoma

I had known for a long time that Steve had only one eye, but we had never actually talked about it. Knowing Steve, I assumed that he probably lost his eye while calf roping or bull riding, but the topic had never come up. While fishing one day, I asked him the fateful question, "Hey Steve, how did you lose your eye?"

"I had cancer when I was a kid," he replied.

"You had **retinoblastoma**?" I asked without missing a beat.

"How did you know that?" Steve laughed. "I usually don't mention retinoblastoma since no one understands what I am talking about."

Retinoblastoma is the most common **malignant neoplasm** of the eye. It most often affects children, usually before the age of 5, and it is very uncommon in adults. The treatment for retinoblastoma has traditionally been surgery, which results in the loss of the eye, but is about 90% successful in completely removing the cancer. Retinoblastoma gained a lot of notoriety in the 1980s and 90s because it was discovered that the mutation of the Rb gene was sufficient to cause the cancer. This concept challenged a previous hypothesis that cancer was caused by the accumulation of a series of mutations, which was known as the **multi-hit hypothesis**. It was thought that on occasion, random mutations would cause normal genes to be changed, so that when these genes were expressed, they would produce abnormal proteins that could contribute to the cell becoming cancerous. These genes in their normal state are called **proto-oncogenes,** and when they become mutated and help to cause cancer, they are called **oncogenes**. We call mutations in proto-oncogenes "gain-of-function" mutations since the protein produced from an oncogene causes the cell to have a trait that it did not have before. An example of a trait gained from the mutation of a proto-oncogene might be continued cell division, even though other cells of the same type have stopped dividing. In a genetic sense, such a mutation is considered **dominant**. This means that although every cell has two copies of each gene, one on each of the **homologous chromosomes**, only one copy of the gene needs to be mutated to provide the cell with the new trait. What was odd about the Rb gene was that it did not require mutation of multiple proto-oncogenes, as mutation of the Rb gene alone was sufficient to cause retinoblastoma.

The study of the Rb gene revealed that it is not a proto-oncogene, but a type of cancer-causing gene called a **tumor suppressor gene**. The protein products of these types of genes normally function to stop cells from dividing, leading to cell division only when it is appropriate. Mutations in tumor suppressor genes are considered to be genetically **recessive**; meaning that the cell must have mutations in both copies of the gene to completely lose the trait that the gene provides. When the trait is lost, this can contribute to the cancer. Therefore, mutations in tumor suppressor genes are recessive "loss-of-function" mutations. The cell no longer has the function of the protein coded by that gene and the corresponding trait is lost; in this case, inhibition of DNA replication. In the case of retinoblastoma, a mutation of both copies of the Rb gene is all that is required to cause cancer, as long as it happens in any cell in the eye generally before the age of 5.

Father

Mother	R	r
R	RR	Rr
R	RR	Rr

Since tumor suppressor genes are recessive and both copies need to be mutated in a single cell before cancer can develop, a person could sustain random Rb mutations in many individual cells without ever having a problem. It turns out that most people that develop retinoblastoma are **carriers**. This means the person would have inherited one mutated Rb **allele,** or copy, so that from birth, all the cells in the carrier's body have one mutated Rb gene. Therefore, carriers have well over a 50% chance of suffering from retinoblastoma because they only need to have a mutation in the other copy of Rb in *any* cell of the eye to develop cancer. Carriers often will have **bilateral** retinoblastoma. Interestingly, the odds of a child born with two normal copies of Rb being diagnosed with retinoblastoma is 1 in 30,000. Why is this? It turns out that it is unlikely to have both copies of the Rb gene mutated in the same cell of the eye in childhood. Almost 50% of all retinoblastoma occurs in families that have carriers; this is called **familial** retinoblastoma. Let's see how this could work. If retinoblastoma is familial, then either the mother or father must be a carrier. In either case, there would be a 50% chance that the carrier would pass the mutated Rb gene on to a child. Geneticists use a **Punnett square** to make this calculation, as illustrated in the figure on the previous page. Letters across the top of the square represent the two copies of the Rb gene from the father. In this case, he is a carrier, as seen by the "R" signifying the normal dominant Rb gene and the "r" signifying the mutated recessive Rb gene. Likewise, the left side represents the mother's two possible contributions. Notice that she is not a carrier. In a Punnett square, each letter simply indicates which gene copy a sperm or egg *could* be carrying. The boxes show the possibilities that will occur when a specific sperm and egg join and combine each gene copy from the mother and father. This Punnett square has four boxes, two of which are "RR," which would be non-carriers, and two of which are "Rr," which would be carriers.

Therefore, from this mating, half of the possible offspring would receive one defective Rb gene. Since well over 50% of those receiving the mutated gene would eventually get retinoblastoma, we can assume that the family of a carrier would have almost half of the children with retinoblastoma.

It turns out Steve did not know much about his retinoblastoma; his cancer occurred before the Rb gene and tumor suppressor genes were discovered. He was aware that retinoblastoma is the most common cause of tumors of the eye, but frankly, he didn't think about it much. Having only one eye has never slowed him down much and none of his friends really notice. However, he is fun to sneak up on if you know his blind side!

Questions about this case:

1. You can figure out with some confidence if Steve's cancer was familial if you ask him some key questions. What would be some good questions to ask Steve?

2. How did the first carrier of a mutant Rb gene occur?

3. In light of the discovery of tumor suppressor genes, is the multi-hit hypothesis still valid?

4. What are some traits that could be gained by a mutated proto-oncogene that would contribute to cancer?

5. In studying evolution, isn't the assumption that most mutations destroy function common? How likely is it to mutate a proto-oncogene to gain a new function (not destroy existing function)? Why is this type of mutation seemingly so common?

6. What is the likelihood of a child inheriting two mutated Rb genes if two carriers had children (you can use a Punnett square).

Questions to go deeper:

1. What is the difference between a benign and a malignant tumor?

2. What are some other treatments for cancer besides the one mentioned here?

Bill and Sara's Visit to the Doctor

A Case Study of Down Syndrome

Bill and Sara were excited. After three years of anticipation, they were finally going to have a baby! They both wanted their first to be a boy, and they already had the name Samuel picked out. Much of the excitement stemmed from the fact that Bill and Sara had a "late start on life." Bill was 43 and Sara was 41. They had been married for nearly seven years, but they were just too busy with work to have kids. Because they had been trying for three years to get pregnant, this news was nothing short of a miracle. They could not wait to be parents! Their excitement quickly turned to fear when their physician called to make another appointment. As it turned out, Sara had a **triple screen** done during her previous visit and the results were abnormal. As they reached the doctor's office, their doctor walked them back to her office and began to explain what the results meant. "Sara, the triple screen simply measures some molecules in your blood that would indicate a higher-than-average chance that your baby has a birth defect," she explained. Their physician continued, "We're going to have to run some tests."

"Why?" they asked.

"Well, the main reason," the doctor replied, "is that we want to test for **Down Syndrome**."

"Oh, no," Sara gasped. "This is because of my age, isn't it?"

"Yes, you are correct," the doctor stated in a somber voice. "Sadly, you have about a 1:100 chance of having a Down Syndrome child."

Nearly 350,000 families live with a Down Syndrome child in the United States (1). On average, Down Syndrome occurs in 1:1,000 live births; however, the frequency increases with age (see Figure 3). Down Syndrome is a genetic disease that leads to a combination of birth defects. This disease was first described in 1866, by a British physician named Langdon Down. The most common symptoms of Down Syndrome are mental retardation, facial abnormalities, and heart problems. Down children also have certain distinguishing features. They are usually very short in stature, have broad skulls, possess long tongues, and they usually have stubby hands with a distinct crease. Down is caused by a **nondisjunction** of **chromosome 21** during **gametogenesis**. You are probably thinking, "What does that mean?" Well, that is exactly what Bill and Sara thought as the doctor began to explain the cause of Down Syndrome, and why the chance of having a child with Down Syndrome increases with age.

Human beings normally have 46 **chromosomes**, which are broken down into 23 pairs. A person normally receives one set of 23 chromosomes from his or her mother, and the other set from his or her father. For example, Bill's cells should contain two copies of chromosome 21, one that came from his father's sperm, and one that came from his mother's egg. These two copies of chromosome 21 are called **homologues**. When a cell's **nucleus** possesses all 46 chromosomes (two sets of 23 chromosomes), it is considered to be **diploid**. However, when it contains only one set of 23 chromosomes, it is considered to be **haploid**. Almost every cell in the body is diploid, except for sperm and eggs (which are haploid). Interestingly, these **gametes** (sperm and eggs) are formed from diploid cells that divide in a very special way. The process by which a diploid cell becomes a haploid cell is called **meiosis** (see Figure 1). If you think about it, having haploid gametes makes sense. What if both the sperm and the egg were diploid? When **fertilization** occurred, the resulting **zygote** would then possess 92 chromosomes (46 x 2). When a cell possesses more than the diploid number of chromosomes, it is called **polyploidy**. This would lead to all kinds of developmental problems that would ultimately lead to death. The process of meiosis works similarly to the regular type of cell division in the body, called **mitosis** (Figure 1).

Mitosis is the process by which a diploid cell divides to form two genetically identical diploid **daughter cells** (Figure 1). Mitosis is constantly occurring in your body, especially in organs like the skin. It is most common during times of growth and during wound healing. For mitosis to begin, a cell needs to replicate the 46 chromosomes, so that it will have enough **DNA** to produce two identical daughter cells. Once the synthesis phase is complete, it will be like the cell has 92 chromosomes (two identical copies of 46 chromosomes), but the copied chromosomes stick together as 46 duplicated chromosomes. The duplicated copies of the chromosomes are called **sister chromatids**. Why is it necessary for the cell to duplicate the chromosomes prior to mitosis? Well, if you think about it, this makes complete sense. If one cell is going to produce two identical daughter cells, it will need to double its contents (including the 46 chromosomes). Therefore, when it divides, two identical diploid cells will be produced. Mitosis begins with a chromosomal condensation. The chromosomes condense to ensure that they do not become tangled up and broken as they are partitioned into the new daughter cells. As they are being condensed, they migrate to the center of the

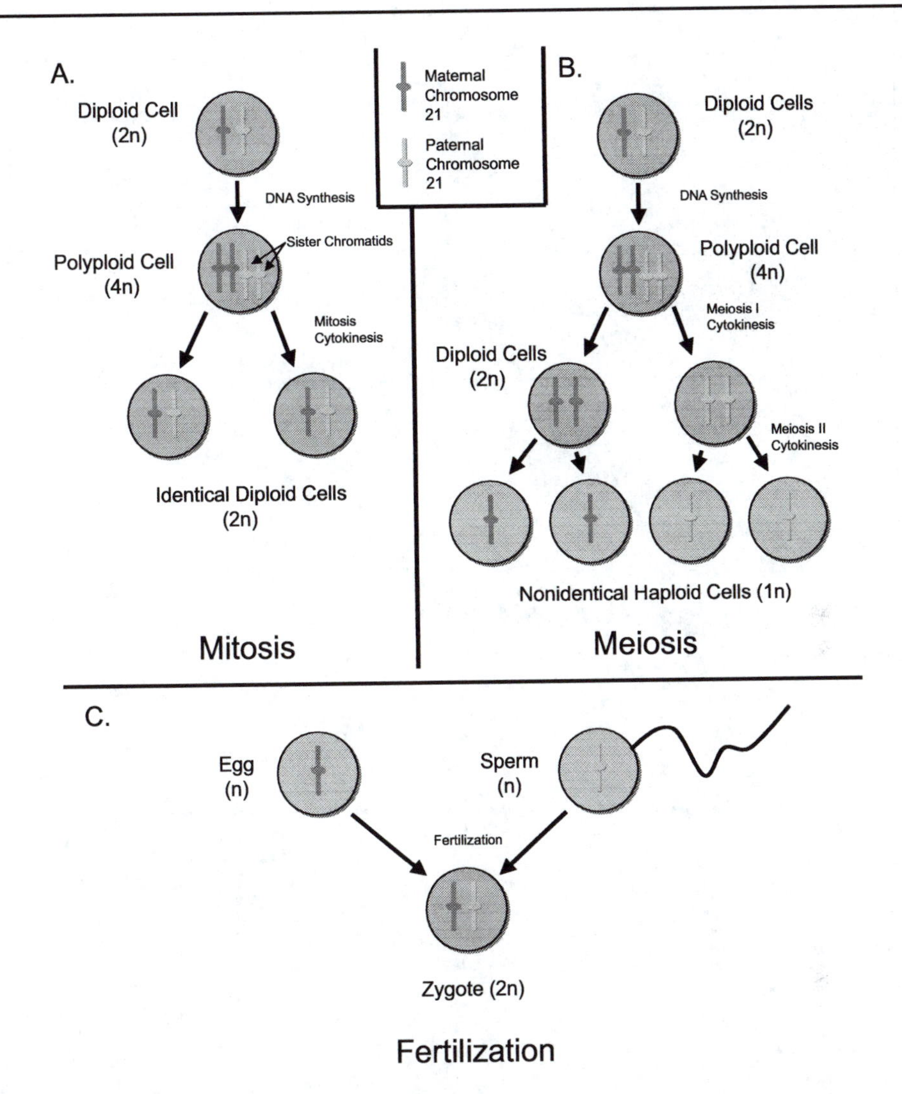

Figure 1: This figure illustrates normal mitosis (Box A), normal meiosis (Box B), and normal fertilation (Box C).

nucleus, and line up so that duplicated chromosomes (sister chromatids) are next to each other (Figure 1). Once this step is complete, the sister chromatids will be pulled apart, and they will eventually end up on opposite sides of the cell. Once this occurs, the cell will split into two new cells, through a process called **cytokinesis.**

The process of meiosis is very similar to that of mitosis. In a very real sense, meiosis is really just two rounds of mitosis placed together without the second duplication of the chromosomes. Like mitosis, meiosis begins with a diploid cell that has dupli-

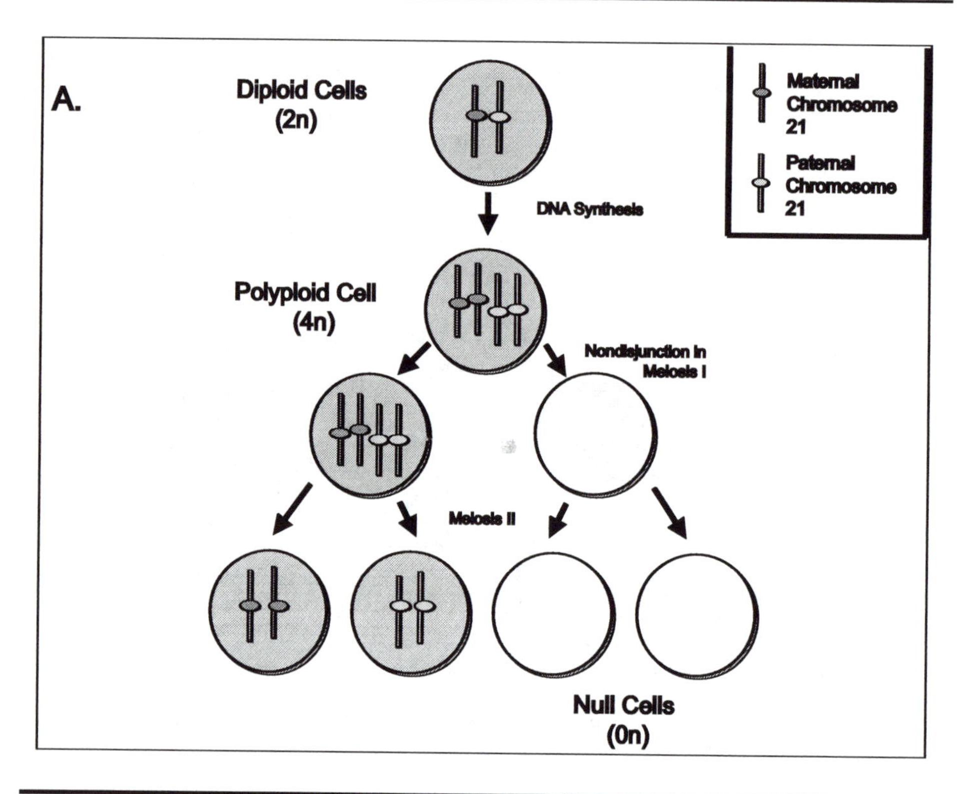

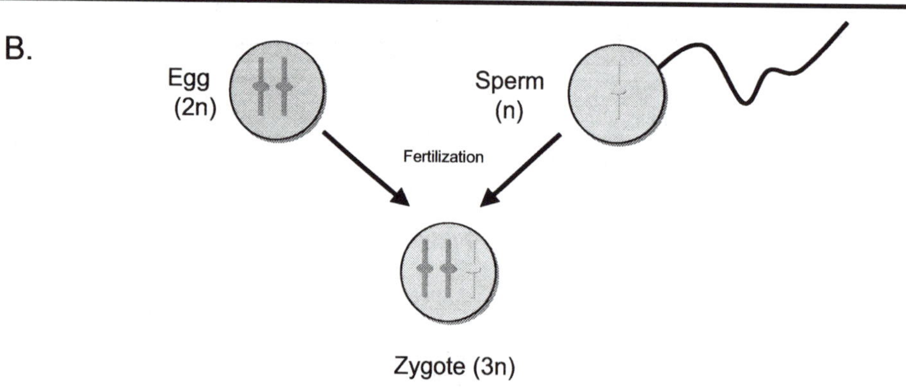

Figure 2 : This figure shows that the product of nondisjunction in meiosis I can result in a fertilized egg with 3 of one type of chromosome, instead of two.

cated its chromosomes; consequently, it has two identical copies of 46 chromosomes. As described for mitosis, the identical chromatids are stuck together as 46 duplicated chromosomes. Once this occurs, the cell will divide into two cells, each containing a set of homologous duplicated chromosomes (23 chromosomes total, but 46 chromatids). Interestingly, one cell does not get all of the mother's homologous chromosomes, and the other, all of the father's. During the first division of meiosis, the homologous chromosomes **assort independently**. This means that each homologous chromosome acts independently. Therefore, the two cells that result from the first division of meiosis can have any combination of homologous chromosomes. If you do the math, this means that there are over a billion different combinations! Once the first division of mitosis is complete, the two resulting cells will then undergo a second round of division. During this division, the sister chromatids for the entire set of chromosomes are pulled apart and are separated into 2 cells. This step ultimately leads to 4 haploid cells (23 chromosomes each; Figure 1).

"This is all great information," Sara replied, "but how does Down Syndrome fit into this?"

"Just bear with me," the doctor answered. "I just want to make sure that you have all of the background information to understand what I'm going to tell you next."

Meiosis is the key process in gametogenesis: the formation of sperm and eggs. In spite of this, the process of **oogenesis** (egg formation) is a little more complicated than "simple meiosis." It turns out that when a female is born, her **ovaries** have already produced every egg that she will ever need. However, these eggs have yet to go through meiosis. The **primary oocytes** are actually trapped in the beginning stages of meiosis, just before the first division, until they are released during **ovulation.** This means that an oocyte can wait up to 55 years to go through meiosis.

"This is why Down Syndrome is more common as mothers get older, isn't it?" Sara asked (Figure 3).

"You are correct," the doctor replied.

As was stated earlier, Down Syndrome is caused by a nondisjunction of chromosome 21 during gametogenesis. Basically, somewhere during one of the two meiotic divisions, a set of chromosome 21s stick together that should normally be separated. This leads to a sperm or egg that has an extra chromosome 21 (Figure 2). When this egg (for example) is united with a normal sperm, the resulting zygote will have 47 chromosomes (three number 21 chromosomes; Figure 2). This is often referred to as **trisomy**. This chromosomal arrangement results in Down Syndrome. In most cases, Down Syndrome is caused by a nondisjunction in the egg.

"I'm sure that you now know why," the doctor stated.

After the doctor finished explaining all of this to Bill and Sara, he asked Sara if she was ready to schedule an appointment to test her baby for Down Syndrome.

"What does it involve?" Sara asked.

"We are going to do a procedure called **amniocentesis**," the doctor replied. "This is where we remove some of the **amniotic fluid** surrounding the baby. Since this fluid

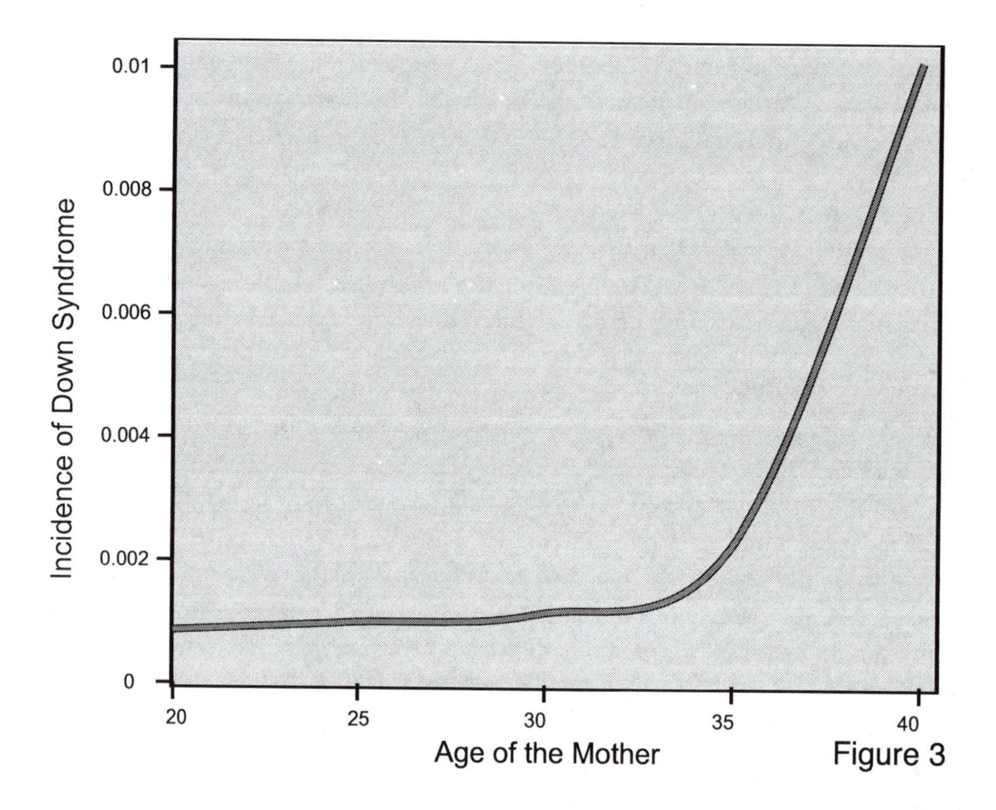

Figure 3

contains some of the baby's cells, we can use them to perform a **karyotype**."

"What is that?" Bill asked.

"Well, this is where we look at the baby's chromosomes to make sure that there is not an extra number 21," the doctor answered.

"Are there any risks associated with this procedure?" Sara quickly asked.

"There is some risk of infection and miscarriage, but this does not occur very frequently," the doctor responded.

"We'll have to think about it," they both replied.

As Bill and Sara left the doctor's office that day, they felt dejected and they had a lot of decisions to make. After doing a lot of research and giving it a lot of thought, they decided that the risk of amniocentesis was just too high. They simply decided to have the baby and hope for the best. Eight months later, they had a normal baby boy. They named him Samuel.

Questions about this case:

1. The stages of mitosis described above have names. List them.
2. What is another example of a disease that is caused by nondisjunction?
3. When nondisjunction occurs, you can produce not only an egg (or sperm) with an extra chromosome, but also an egg that is missing a chromosome. If this egg was fertilized by a normal sperm, what do you think would be the result?

Questions to go deeper:

1. What other birth defects can the "triple screen" detect?
2. If you were Sara or Bill, would you have wanted to know the results of the amniocentesis?

Reference:

1. http://www.marchofdimes.com/professionals/681_1214.asp

Jesse's Accident

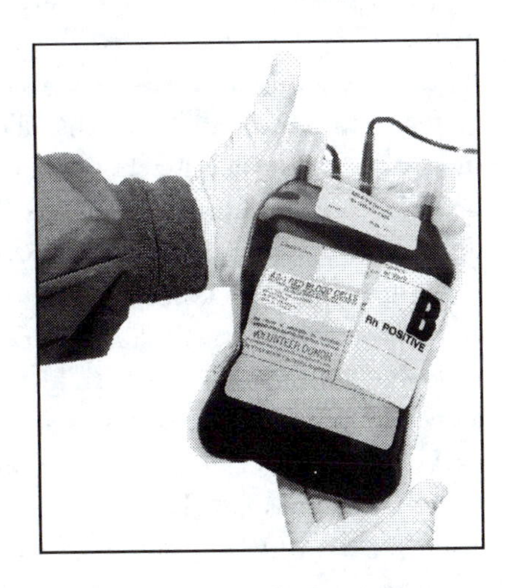

A Case Study of Blood Type

One of the key issues in emergency medicine is the necessity to stop severe bleeding. In many instances, once bleeding is stopped, blood loss has been significant enough to require transfusion of one or more units (pints) of blood. When Jesse was injured in a car accident, the paramedics were concerned primarily with controlling the severe bleeding from a deep wound in his right leg in order to keep him alive. Since the paramedics had no history on Jesse, they infused two units of O negative blood enroute to the hospital. When Jessie arrived at the hospital, the staff was able to contact Jesse's parents, who rushed to the hospital to see their son. When Jesse's parents were interviewed about his medical history, they confirmed that he had an A positive blood type. This information then enabled the physicians to give Jesse additional transfusions of one of four types of blood: A positive, A negative, O positive, or O negative. Because of Jesse's extreme blood loss and the low supply at the **blood bank**, his father volunteered to give blood directly to Jesse. The hospital was happy to receive the donation, but informed Jesse's father that with his AB negative blood type, his blood could not be given directly to Jesse. When he pressed for an answer about why he could not donate to his son, the physician made it clear that the donation to the blood bank was important. However, the answer to why a father could not donate blood directly to his son would require an understanding of basic immunology and genetics.

Blood type is determined by two major sets of molecules on the surfaces of blood cells. The first set is called the **ABO antigens**, and the second is called the **Rh factors**. These, like all proteins, have the instructions for their production within the cell's DNA. Human DNA is held within 46 chromosomes that exist as 23 pairs. The genetic information for the ABO blood group is on chromosome 9 and since there is a pair of chromosome 9s, there are two sets of instructions on what molecules to make. While humans have two separate copies of the information that makes the ABO blood **antigens**, there are actually three different gene types or **alleles** that could be found there. The three possible alleles that can be found at this position on chromosome 9 are an allele for the A antigen, an allele for the B antigen, and an allele that does not produce an antigen, so it is called O. A person will have some combination of these three alleles on his or her two gene copies found on the number 9 chromosomes. For example, Jesse's dad has an AB blood type. Therefore, one of Jesse's dad's chromosome 9 has the A allele and the other has the B allele. Because of this, he will make both A and B antigens, so his blood type is AB. In this case, the A and B alleles would be considered **codominant** because when both are present, they both show up in the physical expression. Jesse, on the other hand, has A type blood, so one of his chromosomes has the A allele and the other has either has another A or an O allele. If his two chromosomes both have the A allele, then he is called a **homozygote**, but if he has one A allele and one O allele, he still makes the A antigen and has type A blood, but is called a **heterozygote**. Type O blood would occur only if both alleles are O genes and would therefore be considered **recessive**. The recessive gene does not affect the physical appearance unless a dominant gene is NOT present. So, genetically speaking, when we talk about the physical appearance of a genetic trait, it would be called the **phenotype,** and when we talk about the genetic makeup of that trait, it would be called the **genotype**. The use of the phrase "physical appearance" can mean what exists on a molecular level (presence of a B protein on blood cells), as well as what can be seen with your eyes (i.e., blond hair). Therefore, if Jesse happened to be a heterozygote, he would have a phenotype of A type blood with a genotype of heterozygous AO blood type.

The Rh factors (or antigens) are less complicated than the ABO blood type. Humans again have two copies of the Rh gene (since chromosomes come in pairs), but only two possible alleles exist, Rh+ or Rh-. The dominant Rh+ gene codes for the Rh antigen, and the recessive Rh- gene codes for nothing. Like the ABO system, when either chromosome carries the + gene, then the blood type is +. The only way to be Rh- is to have both alleles be Rh-. Therefore, a person who has Rh- blood must be a homozygote, since both alleles would need to be Rh-. The Rh+ person could either be homozygous dominant or heterozygous.

Even with this explanation, Jesse's dad was still left wondering why his son has a different blood type than he does, and why he can't give blood to his son. First let's deal with why he can't give blood to a person of different type. The human immune system is fine-tuned to differentiate between molecules that belong to the body and those that are foreign. The goal is to ignore the "self" molecules and attack and eliminate the foreign molecules (and the associated foreign organisms). A human with A type blood (Jesse) has an immune system that has never encountered the B antigen and would consider it foreign. If Jesse were given AB blood, then Jesse's immune system would find the B antigens and form **antibodies** against them. This would result in the death of all AB type blood cells. This is a great way to protect against invading organisms like viruses, but makes blood typing important. The Rh factors work the same way. Jesse has Rh+ blood, so his immune system does not consider the Rh+ antigen foreign. Whether Jesse receives a transfusion of Rh+ blood (the presence of the antigen) or Rh- blood (the absence of the antigen), Jesse's immune system will not be compelled to attack the cells.

Why then does Jesse have a blood type that is incompatible with his father's? The answer has everything to do with genetics and Jesse's mother's blood type. Jesse's dad is type AB. When **sperm** get made, they only get one of the two chromosome 9s found in the body, so half of the sperm that Jesse's dad makes will have a chromosome 9 with an A allele, and the other half will have the B allele. Because Jesse received one of his chromosome 9s from his dad, that means that Jesse got his other chromosome 9 from his mom. Since Jesse has A type blood, he does not carry the B gene; therefore, he is either a AA homozygote or an AO heterozygote. Consequently, he must have been the result of fertilization with a sperm with an A allele. What is Jesse's mom's blood type? From this information we cannot be sure. Jesse got an A allele from his dad and either an A or O allele from his mom. It is then possible that his mom has type B blood (heterozygote **BO**), **AB** blood, A blood (heterozygote **AO** or homozygote **AA**), or type O (homozygote **OO**). Geneticists use a **Punnett square** to analyze these types of problems. The possible genotypes for Jesse's parents are aligned on the sides of a square to represent how alleles might be donated when sperm and **ovum** are made, and then fertilized. Notice the five scenarios in the figure. Dad will always be AB blood type, and there is always at least a 25% chance in having a type A child. If we were to perform similar Punnett squares for Jesse's Rh+ blood type, we would find that his mother must be Rh+. Jesse's dad is Rh- (must be a homozygote), so for Jesse to receive an Rh+ allele, it must come from his mother. But, we do not know if she is homozygote (+/+) or heterozygote (+/-) since both have at least one Rh+ allele that could be donated.

A good illustration of the importance of understanding Rh blood typing is that of Rh incompatibility during pregnancy or "**Rh disease**." When an Rh- mother gives birth to an Rh+ child (with an Rh+ father) the process of birth will expose the mother to the baby's blood and Rh+ antigen for the first time. This will serve as an immunization of sorts. Her immune system will use **B-cells** to produce antibodies against the Rh+ antigen. It is important to note that these B-cells have no relationship to the B type blood antigen. Some of the B-cells involved in this immune response would

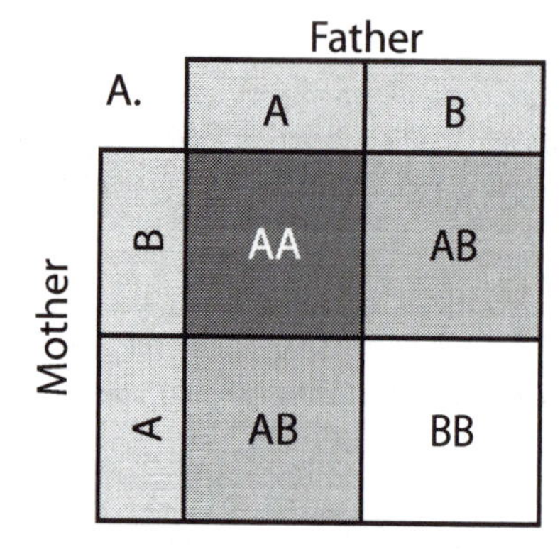

B.

Mother \ Father	A	B
A	AA	AB
A	AA	AB

C.

Mother \ Father	A	B
A	AA	AB
O	AO	BO

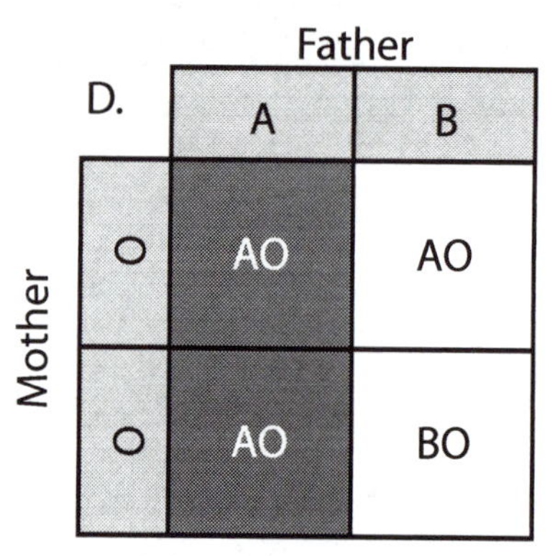

E.

Mother \ Father	A	B
B	AB	BB
O	AO	BO

then be stored away as memory B-cells so that the foreign molecules can be quickly eliminated if they are identified again. If this mother conceives a second Rh+ child, the mother's B-cells would not encounter the fetus's blood because the **placenta** allows only molecules to pass back and forth (cells are too big). Unfortunately, the antibodies that these B-cells make are small enough to pass across the placenta and attach to the Rh+ antigens on the fetus's blood cells. This would result in their destruction.

The best treatment for Rh disease is prevention. When an Rh- mother delivers an Rh+ child, the mother can be given purified antibody against the Rh+ antigen. The purified antibody will locate the Rh+ antigens in the mother's blood, bind to them, and in doing so target them for elimination by the mother's immune system. This would happen relatively quickly so the mother's B-cells would not have time to identify the Rh+ antigen and make antibodies against it. This treatment would need to be repeated after each Rh+ baby an Rh- mother delivers.

Most people do not fully understand the importance of blood banks and blood donations until they have a loved one in the hospital in need of one or more units of blood. We often cannot give blood directly to our loved ones because of issues of blood type (and even if we could, they often need much more than we can give). Therefore, we depend on anonymous donors to supply blood when our loved ones need it. The problem is this: only about 5% of eligible people actually donate, while almost 95% of people will need blood sometime during their lifetime (1). Donating blood is really very simple; it does not take much time, and it is not very painful (although many people do not believe that). If you donate through the Red Cross, you will be asked a number of questions regarding your medical and personal history to ensure that it is safe for you to donate and that the blood you donate is safe for others to receive. After your donation, chances are good that you will receive a donor card that will indicate the number of units (pints) you have donated over the years and your blood type. Donations usually skyrocket after major catastrophic events, as they did after 9/11, but before long, donations decline back to dangerously low levels. Blood banks routinely run low on donations and yet we all assume it will be there when we need it. I am particularly proud of my father in this regard. He does not make a big deal about how often he gives blood, but I found a "10 gallon" pin on his dresser. The pin represents 80 donations.

Questions about this case:

1. Why did the paramedic give Jesse O- blood before they knew his blood type?
2. What blood type would allow the receipt of any other type as a transfusion? Why?
3. What might be concluded if you found that Jesse's mother's blood type is B+ and her genotype is homozygous BB?
4. Write out the Punnett squares for Rh factor for Jesse's parents.
5. Why is donated blood extensively tested before use?

Questions to go deeper:

1. What would be a treatment for Rh disease if the mother was not given the preventative treatment?

Reference:

1. www.redcross.org

The Boy in the Plastic Bubble

A Case Study of Gene Therapy

Bob and Linda had no idea of the ramifications of the news that they just received. Their infant son, Luke, had been sick almost since birth. It seemed like he would get one infection after another, and the latest bout with pneumonia nearly killed him. Weeks of testing finally provided the answer: Luke was **adenosine deaminase** (ADA) deficient. It seemed like a harmless description of a **congenital disease**. Their son was only missing a single **gene**. How could this cause all these problems? But, as the doctors continued to explain, Bob and Linda realized that this could be a life-threatening disease.

ADA is simply an **enzyme** that helps to make molecules required in manufacturing **DNA**. While the action of this enzyme is very important, it's not the lack of the **product** that causes the problem: it is the excess of **substrate** that builds up over time. The molecule that ADA consumes in this reaction is particularly toxic to immature **T lymphocytes.** These are white blood cells that are the key for immune system function. Therefore, since Luke is unable to process the ADA substrate, the toxic buildup is killing his immature T-cells. T-cells mature into two varieties, including **helper T-cells** and **killer T-cells**. While losing killer T-cells can have serious ramifications, losing the helper T-cells is catastrophic. Helper T-cells are the cells that send molecular messages to tell most of the other cells in the immune system what to do. Without helper T-cells, even the most harmless virus cannot be identified and killed. Ultimately, if mature helper T-cells are not produced, Luke has absolutely no immune system. As a result, the most insignificant infection, like the common cold, could kill him. The disease that results from this problem is called **severe combined immune deficiency** (SCID). People also refer to this disease as the "bubble boy disease." This is because there was a famous instance of a boy with SCID who lived for 12 years in a sterile plastic bubble.

Bob and Linda were concerned about how this could have happened. They wanted to know what they could have done to prevent it. Luke's specialist assured them that they did nothing to cause the disease, and there was really no way that they could have prevented it. The specialist then went on to explain the basic biology involved in SCID to Bob and Linda. We all have 46 **chromosomes,** which are arranged in 23 pairs. Each pair provides the information on how to make the proteins for a specific set of traits. While each member of the pair codes for the same set of traits, they are not identical. For each pair, one chromosome is inherited from the mother and one is inherited from the father. So for something like the ADA gene, which is found on chromosome 20, the information on both chromosomes should be the same, since that trait shouldn't vary from person to person. Conversely, the genes that code for something like eye color do vary. They are found on the same chromosome pair (one from mom, and one from dad), but the specific genes on each chromosome probably contain slightly different instructions for making **protein**. Therefore, traits are simply a function of which proteins are being produced. Ultimately, the information carried by DNA dictates the amino acid sequence for each protein that cells could make. In addition, DNA carries the information regarding which proteins are produced and when, where, and how much of each is made. A chromosome is just a long chain of DNA that is organized into genes. A gene is a section of the chromosome that carries the information to produce a single **amino acid** chain that leads to a single **protein**. Chromosomes are located in the **nucleus** of a cell. Consequently, for a protein to be made, the gene that codes for that protein must be copied and sent out to the **cytoplasm**. That copy is called a messenger RNA (**mRNA**) and the copying process of making mRNA is called **transcription**. Once in the cytoplasm, the mRNA will be used in a process called translation to build the appropriate protein. Normally, a person carries two normal copies of the ADA gene, one on each chromosome 20. As ADA is needed, the ADA gene is transcribed and translated: this process leads to functional ADA protein. In Luke's case, both of his ADA genes are **mutated**. This means that he has one or more errors in the DNA sequence that codes for each of his ADA genes. This results in ADA protein that does not function. Where did these mutations come from? There are a number of possibilities. It is possible that both Bob and Linda carry one mutated copy of the ADA gene. If this is the case, they would not be aware of it. Because they would each still have one good copy of the ADA gene, this would allow for the production of enough normal ADA that they would appear normal. Luke then would have inherited the mutated chromosome from both mom and dad. Mutations occur all the time. Lots of different drugs, chemicals, and environmental factors can cause them. Sometimes they are just spontaneous errors that are made when cells copy DNA.

Now that Bob and Linda had a handle on what had happened, they expressed concern for what this meant for Luke. While the plastic bubble idea was ingenious, Luke's specialist explained that there are now better alternatives. Scientists can purify ADA protein from cows and inject that into a patient with the deficiency. The maturing T-cells would then absorb enough of the bovine ADA to mature. Therefore, patients can be kept relatively healthy. **Recombinant** ADA can also be used. This process entails placing a **cloned** ADA gene (simply a genetically identical copy) into a bacterium and then growing large volumes of the bacteria. The **rADA** is then purified from the bacteria and injected into the patient. Of course, the problem is that the patient must have repeated injections. The best solution has been **bone marrow transplants**. Since the immature T-cells are produced by the bone marrow, a transplant provides the patient with a marrow that makes T-cells that produce functional ADA. Since the patient has no immune system, the transplant cannot be rejected. However, a suitable donor must still be found since the new immune system supplied by the new bone marrow could attack the recipient's cells (this would be called **graft versus host disease**).

In the following months and years, there was a problem finding a suitable marrow donor for Luke; he was relying on ADA injections. After seeing another specialist, Luke's parents were asked if they wanted to try a revolutionary treatment called **gene therapy**. In this treatment, a healthy version of the ADA gene is introduced into

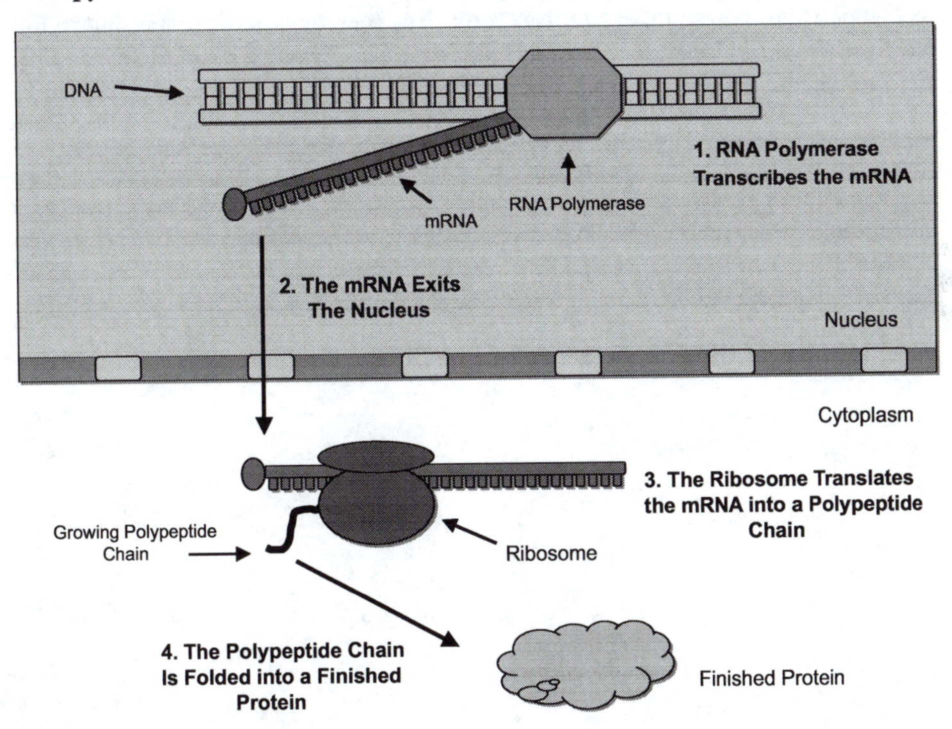

This figure illustrates the steps required to produce a protein from the DNA code.

Luke's own bone marrow cells. The hope is that marrow cells that are responsible for making T-cells will receive some of the healthy gene and use it to make ADA. If this worked and these cells established themselves back in the boy's bone marrow, then he would be normal and would no longer require injections. Gene therapy of this nature requires that scientists would have already cloned the ADA gene. Scientists would then have to figure out how to take this cloned gene and engineer it into a **virus** that could infect Luke's bone marrow cells without harming them. This way, they could take some of the boy's bone marrow, infect it with the engineered virus, and inject it back into Luke. The goal of this technique is to have the virus infect the bone marrow cells and have the viral DNA mix with Luke's DNA. If the technique works, then the new bone marrow cells will re-populate the bone marrow and begin producing T-cells that make ADA. It sounds so easy. Gene therapy has been attempted on a number of diseases and the techniques vary. The basic goal of this therapy is to get a specific gene expressed in the cells that need it, so that the symptoms of a specific disease can be prevented. This is why scientists would work to get the ADA gene into the cells that produce mature T-cells. Although this sounds easy, this can be quite a difficult task, depending on what type of cells need the gene in question. The amount of the gene product needed can also present a problem.

The decision to have Luke go through the gene therapy procedure was quite a big one for Bob and Linda: one that would require a lot of thought.

Questions about this case:

1. What would be the likelihood of Bob and Linda having a child with SCID if they both carry one mutated ADA gene?
2. Will the rADA work just as well as the ADA in treating SCID?
3. If both of Luke's ADA genes are mutated, do his cells still transcribe and translate those genes?
4. If mutations happen frequently, why are diseases like SCID so rare?
5. What would be the potential problems with gene therapy for a disease like cystic fibrosis?

Questions to go deeper:

1. How is it decided which genes are transcribed in a given cell?
2. What could be an ethical dilemma regarding gene therapy?
3. What would be some of the benefits, drawbacks, and ethical issues of using this technology to alter plants?
4. How does a bone marrow transplant work?
5. How is SCIDs like AIDS?

Reference:

1. http://users.rcn.com/jkimball.ma.ultranet/BiologyPages/G/GeneTherapy.html

Convicted by a Paper Cup!

A Case Study of Convicting a Rapist with DNA Evidence

This is the type of case that police investigators both love and hate. Rape is tragic. It is so awful that some victims wish that they hadn't survived. Yet, sending a rapist to prison brings the sense of accomplishment that investigators live for. This is very much that type of case. Megan was lucky to be alive. She was badly beaten during the rape, and she did not get a good look at the perpetrator. She could give the police only a rough description of the assailant's height, weight, and hair color. Megan was smart enough to call the police immediately and at the hospital the medical personnel collected a semen sample from Megan's vagina. The police also collected another sample from her clothes. Sadly, without a good physical description, the police simply did not have much to go on.

Joe Stevens was the lead detective on the case. The investigation began with three parts: compare the physical description that Megan provided with those of previously convicted sexual felons in the area, run a **DNA fingerprint** on the semen samples and check the results against the existing database, and interview anyone in the area of the rape who might have witnessed something. As the investigation progressed, there were some leads, but there were many setbacks. Sadly, the DNA fingerprint did not match anything in the database. However, the minimal physical description given by Megan was also given by a witness. This description loosely matched a few local men, who had prior sexual offenses. Interviews with these suspects and police line-ups with Megan and the witness provided further leads, but nothing concrete. All of the suspects had good alibis, except for one, whose alibi was weak at best. During interviews with this suspect, there where clear inconsistencies with his testimony. However, without a positive identification from Megan or positive **DNA evidence**, there would be no case.

DNA evidence was their only hope, since Megan did not see enough of the assailant to make a positive identification. But, what is DNA evidence? Where does it come from? How reliable is it? In order to answer these questions, we must discuss what DNA looks like. DNA is a **polymer** of molecules called **nucleotides**. Nucleotides are composed of **sugar**, a **phosphate** group, and one of four different **bases**. Polymers of nucleotides are formed by connecting the sugar of one nucleotide to the phosphate of the next. This linkage creates a repeating sugar-phosphate chain that has the bases attached to the sugars. In DNA, the sugar is called **deoxyribose**. Therefore, DNA gets its name from an abbreviation of "**d**eoxyribo**n**ucleic **a**cid." The bases are called **adenine**, **guanine**, **cytosine**, and **thymine** (we usually abbreviate them with their first letters). Since the bases come in four different types, the order of the bases within the polymer will carry a "code." Ultimately, the code will carry the information to guide how and when specific proteins are to be made. DNA is double stranded: meaning that two of the polymer strands, described above, will stick to each other by virtue of **hydrogen bonding** between the bases. The chemical structures of the bases allow adenine and thymine to bond to each other. Cytosine and guanine bind to each other as well. This is something like a ladder, with the outside rails being the alternating sugars and phosphates and the rungs being the pairs of bases. Some of the sequences of these bases are organized into regions that directly code for the amino acid sequence of a protein. These regions are called **genes**.

How does this relate to DNA evidence? Humans have 46 **chromosomes** that exist as 23 pairs. We number the different chromosomes 1 through 23, and each person normally has two of each chromosome: one that is inherited from the mother and one from the father. One set of 23 chromosomes contains the genetic code to produce all of the proteins that a person would need. However, the traits we see in a person are the result of two complete sets of coding information. Since all humans have a large number of genes that are exactly the same (for example, normally we all make the same proteins for **glucose metabolism**), there are many chromosomal parts that are the same from person to person. But, there are also large sequences of DNA that exist between genes that do not code for proteins. These sequences can be **mutated** over time, and because they do not represent the code for a protein, the mutation is not detrimental. It turns out that these sequences have changed so much over time that scientists have shown that many of these sequences can be used as a "fingerprint."

In order for a forensic scientist to obtain a DNA fingerprint, he or she needs some source of DNA. This DNA can come from any part of the human body that contains cells, and, therefore, DNA. Often a very small sample is recovered, like a single hair or a dry blood drop that contains only a tiny amount of DNA. A sample like this is usually subjected to a technique called **polymerase chain reaction** (PCR). This technique allows scientists to amplify a specific small part of a chromosome so that enough DNA can be used for the fingerprint. PCR is used to amplify a section of DNA that is known to be nearly unique from person to person. The amplified DNA is then be cut into pieces using molecular scissors called **restriction endonucleases**. These enzymes cut DNA only when they identify specific nucleotide sequences, called **recognition sites**. There are hundreds of different restriction enzymes, and each one cuts at a different recognition site. Since the DNA the scientist is working with is nearly unique from person to person, the recognition sites for these enzymes differ in location and number from person to person. The resulting DNA pieces are then separated by size and visualized using a procedure called **electrophoresis**. If the portion of the human chromosome that was amplified is truly unique from one person to another, then the number and location of restriction endonuclease recognition sites should be unique. It would follow that the number and size of the resulting DNA pieces would be unique as well.

Basically, DNA evidence is simply a comparison between DNA fragments taken from cells found at a crime scene and cells from a suspect. The problem is that the suspect in Megan's case cannot be forced, without a court order, to give a DNA sample. Interestingly, he declined to give one voluntarily. This is the type of problem that detectives hate. Joe Stevens believed that he had the rapist in custody, but he could not prove it. Joe decided to do something risky. He told the suspect that he was free to go, but then he had a detective secretly follow him everywhere he went. It didn't take long. While eating in a fast-food restaurant the suspect ordered a soda and drank it through a straw. After he was finished eating, he left the soda on the table! As the suspect left, the detective donned a rubber glove, picked up the cup, and returned with the evidence. Enough DNA was found in the saliva on the straw that PCR could be performed. The fingerprint was then compared to Megan's DNA, DNA from the semen sample taken from Megan's clothes, and DNA from the vaginal semen sample. In our favorite crime TV shows, this evidence is processed in a matter of minutes. However, in reality, it takes days. Joe and his fellow detectives were anxious, but they finally got the news they were waiting for. Joe took great pleasure in personally arresting the suspect.

In court, the DNA evidence was clearly the most damning. Prosecuting attorneys showed pictures of the sizes and numbers of the DNA pieces and no one could dispute that the semen samples matched the suspect's DNA. The use of DNA evidence in such cases has shown to be tremendously reliable. Therefore, the defense attorney could only argue that the evidence was tampered with, which he did, to no avail. This suspect was found guilty as charged. This is the part of the job Joe loves; catching the bad guys and putting them behind bars. Megan will never fully recover, but seeing the police catch her rapist and convict him was a positive step.

Questions about this case:

1. Would this trial be any different if we found out that our suspect had an identical twin?
2. If a sample of old dried blood works in a fingerprint, just how old can a sample be?
3. Why would they include Megan's DNA in the fingerprint analysis?
4. What are some of the other reasons a DNA fingerprint might be used?
5. Where does DNA in saliva come from?

Questions to go deeper:

1. In order to protect individual rights, a person cannot be forced to give a DNA sample. Do you agree with this? How would you feel about making it significantly easier to get a suspect's DNA?
2. How would a fingerprint be used in a paternity test? Would the father and the child have the same exact DNA?
3. What if the fingerprint of the suspect matches the vaginal semen sample, but not the sample from the clothes?
4. Name a time when a person does not have 46 chromosomes.
5. Recently we have discovered "human chimeras." What are these, and how do they affect the use of DNA evidence?

This Cat Really Might Have 9 Lives

A Case Study of Animal Cloning

Jill's cat, named Boxer, is her best friend, but he has been getting old. Recently, Jill has been struggling with what life will be like without Boxer, but last week she came upon information that indicated that she may never have to know what that would be like. Jill found out that with the current advances in cellular research, companies are claming that they can **clone** Boxer after he passes. In order to do this, these companies claim that **cryopreservation** of a small skin snip is all that is needed in advance to prepare for the day that Boxer dies. She is baffled. "How is this possible?" she thinks. Would she really get her same Boxer back? She can't believe that it could be that easy. She needed some answers, so she went to a close friend who happened to be a physician. "Well," Dr. Brady began, "I understand your concerns. First, I must say that it is possible to clone Boxer, but I need to add that your concern about whether or not you will have a cat identical to Boxer is valid. To understand why, you need to know something about animal cloning techniques." Jill listened intently as Dr. Brady began to explain how animal cloning works.

Ultimately, Jill would like to have a cat born that is a genetically identical copy of Boxer. To do this, some of Boxer's living tissue must be cryopreserved before he dies. This simply involves freezing a small skin snip through a special process that does not kill the cells. The cells are then stored indefinitely in a small vial submerged in liquid nitrogen (-195°C or -320°F). The cloning process is relatively simple, as well. A **mature ovum** is taken from a donor female cat and placed in a dish along with Boxer's carefully thawed skin snip. Using tools that can be manipulated on a microscopic level, the **nucleus** of the **unfertilized** ovum is removed and replaced with a nucleus from a cell in Boxer's skin snip. This new hybrid cell acts now much like a **fertilized egg** that is genetically identical to Boxer. This cell is then placed in a female cat, which will hopefully carry it until a kitten is delivered.

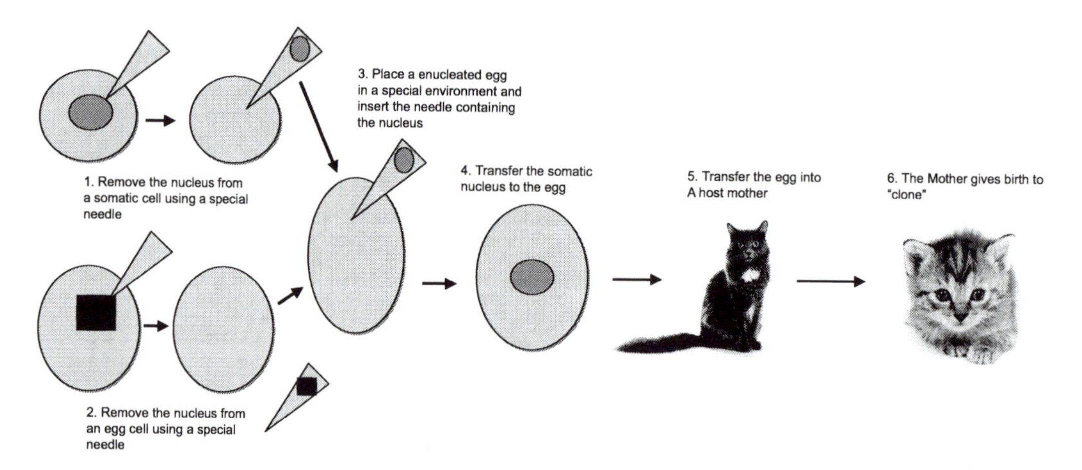

Normal domestic cat cells have 38 **chromosomes**, 19 matched pairs. When cats make **sex cells** (sperm or ovum), they go through a process called **meiosis** where cells are made that have only 19 chromosomes. These sex cells only have one copy of each chromosome, so they are called **haploid**. When fertilization occurs, two haploid cells (a sperm and an egg) join to make a cell with 38 chromosomes, or a **diploid** cell. So, if we simply take an unfertilized ovum and give it a diploid nucleus (from Boxer) then it is similar to fertilizing it. This ovum should then begin to divide like a freshly fertilized ovum and develop into a cat. This process is called **somatic cell nuclear transfer** (SCNT). This makes sense as long as you know that any cell that is not a sex cell is considered to be a **somatic cell**. Therefore, a somatic cell from Boxer's skin snip would be used to recover a nucleus that would be placed in the donor egg. This diploid cell would then be placed into a female cat to carry until birth. Is this process really that simple? The answer is no. Removing a nucleus from an egg and replacing it with a new one is quite disruptive, as you might imagine. Multiple eggs (thousands) are processed in the SCNT technique in order to get a viable pregnancy. Also, since this technology is so new, it is not known how "normal" the cloned animal really is. A good example is the first cloned animal, which was a sheep named Dolly. Dolly had a number of medical problems, including premature arthritis. However, we cannot be certain if the cloning procedure was the cause.

Of course, there has been controversy surrounding this technique. In this particular case, we are simply creating a cloned animal, a genetically identical copy. In a very real sense, this is just like making an identical twin. This might be completely acceptable with Jill as long as Boxer II was healthy, but what about the use of SCNT for human cloning? As you probably know, this is a heavily debated question. There are two parts to this question that should be addressed here: first, human cloning for purposes of creating adults, and second, human cloning for purposes of creating

embryos. Cloning for the purpose of creating an adult has yet to be attempted, is highly controversial, and is illegal in many countries. Most scientists agree that cloning adult humans is not an appropriate use of the SCNT technology and is currently prohibited in labs using U.S. government funding. On the other hand, cloning for the purpose of creating embryos, also called **therapeutic cloning**, involves the use of SCNT in order to produce **stem cells** for use in research and potential therapeutic treatment. This type of cloning is quite common, and the work has been published in reputable scientific journals. This type of cloning is also very controversial. Therapeutic cloning uses the same process as that used to clone an adult, but instead of placing the egg with the new nucleus into a surrogate mother, the eggs are grown in a dish. After the eggs develop for a short time, they are used to harvest stem cells. In the process, the eggs are destroyed.

You may ask why scientists would go through all this trouble for some stem cells. Good question! The answer lies in what stem cells are and in what they do. It is quite an amazing process to have a single egg develop into a body, consisting of billions of cells of hundreds of different types. Since the human body develops from a single egg, this means that the initial egg has the potential to produce all of these different types of cells. The process by which an egg produces these different cell types is called **differentiation**. In order for this to occur, the egg will first divide into 2 cells, and then into 4, and then into 8, and so on. As this happens, the cells receive and produce signals so that they begin to differentiate. At first, they make three different types of cells, but then these three types continue to differentiate into the many different cell types of the human body. The undifferentiated cells in that initial embryo are called **totipotent stem cells**. Cells after the first rounds of differentiation are called **multipotent stem cells**. These cells are named according to how much potential they have to differentiate. There are also stem cells that can be recovered from umbilical cord tissue and from bone marrow that still have some ability to differentiate. These cells are called **adult stem cells.** You may be thinking, "So what? How can scientists use a stem cell's ability to differentiate, anyway?" Well, if you have a disease like **Alzheimer's**, stem cells may be your only hope.

For many years, scientists have known that once certain cell types differentiate, they cannot revert back to the undifferentiated cell they came from. For example, if a neuron in the brain dies for some unknown reason, the differentiated neuron next to it cannot revert back into a dividing cell in order to replace the dead neuron. This can present a major problem when we have a disease or an injury in a tissue that is fully differentiated and cells can no longer divide (like in the nervous system or the heart). In these cases, there is no way for the body to replace the diseased or damaged cells; therefore, the damage is permanent. This is why many severe spinal cord injuries and damage from heart attacks is irreversible. The hope of scientists who study stem cells is to use them in situations like these and in many others. Potentially, stem cells could be given the correct molecular signals and environment to differentiate into adult cells that could replace those that are diseased or damaged. The list of diseases that scientists hope to treat ranges from Alzheimer's to spinal cord injuries. However, currently, these treatments are only in the research stages. Recently, scientists have shown that some stem cells can have their differentiation reversed. This discovery has opened up the possibilities of performing research on adult stem cells without having to do SCNT to get them.

Questions about this case:

1. In your opinion, how different would Boxer's clone be from Boxer?
2. How would performing SCNT be different than doing *in vitro* fertilization?
3. What are some ethical reasons cited by people who are against human cloning to produce adults?
4. What are some ethical reasons cited by people who are against human cloning to produce stem cells?
5. If scientists are able to reverse the differentiation of adult stem cells, what would be some of the possible ramifications?

Questions to go deeper:

1. What do you think about using surplus *in vitro* fertilized embryos as a source of totipotent stem cells?

Observations on the Galapagos Islands

A Case Study of Charles Darwin and Natural Selection

Few people have so profoundly affected their field of study as Charles Darwin affected the field of biology. Darwin's seminal work, entitled *On the Origin of Species,* was published in 1859 and has been the object of much research and debate ever since. Darwin's ideas were borne from a series of **observations** that begged a rational explanation. Many years of gathering data and refining Darwin's ideas have resulted in the **Theory of Evolution.** To make sense of Darwin's work and the resulting Theory of Evolution, it is important to take a look at his travels and original observations.

Charles Darwin was born in 1809 to parents who had hopes that their son would be a physician. By age 22, Darwin had instead earned a theology degree from Cambridge; yet he still did not feel like he had found his calling in life. His real desire was to be out studying nature. With the help of a friend, Darwin was appointed to be the naturalist on a ship called the *Beagle* that was to make a 5-year journey around the world. In particular, the captain of the *Beagle* was to chart the coastline of South America. During both times at sea and in the numerous stops on the South American coast, Darwin collected many samples. Darwin also kept extensive notes regarding how the animals and plants that he observed were particularly well adapted to their specific environment. As the *Beagle* departed the South American coast, it made a stop at the Galapagos Islands, which are off the coast of present-day Ecuador. On this set of islands, Darwin made some of his most formative observations, and solidified his thoughts on evolution.

What did Darwin observe that made such an impact on his thinking? To the untrained eye, many of these observations may seem unexciting, but when they are fit into an overall picture, they become very enlightening. Darwin noticed that the Galapagos Islands were home to a variety of animal species that were found only on the islands. However, each species closely resembled animals on the mainland 600 miles away. In addition, while studying the **fossils** on the South American mainland, he noticed that these extinct animals most closely resembled the animals in that local area, rather than animals found elsewhere in the world. Darwin reasoned that what he saw supported **ancestry**. Darwin then **hypothesized** that the animals he saw on the mainland represented "descent with modification" from the extinct fossilized animals. Furthermore, he argued that since the volcanically formed Galapagos were so geologically young, they could not have had any original inhabitants. Therefore, the animals on the Galapagos must have been ancestors of the similar animals on the adjacent mainland. The observation of structural **homology** between different organisms also supported ancestry. Fossils revealed to Darwin that extinct animals had homologous limb structures to living animals, thus indicating relatedness. This idea of limb homology has carried over into more current studies of molecular biology, including the study of protein and gene homologies. Although the idea of ancestry seemed realistic, Darwin still needed to explain how these species **evolved** into new species over time.

As Darwin continued to observe plants and animals, he noticed that **populations** of specific organisms had the ability to increase in numbers as generations passed, but would at some point be limited by the environment. When the environment became a limiting factor, **competition** for resources would occur. As a result, the individuals who were best **adapted** to their changing environment would have the best chance to successfully reproduce. Darwin explained that the individual who was best adapted to compete for dwindling resources would reproduce the most. This would ensure the genes that code for those "best-adapted" traits would be passed along to the next generation. Population biologists would say that this organism was "**successful**" because its genes remained in the **gene pool**. Therefore, as time passed, the traits that favored survival would become more common in the population. This implied that the **alleles** that code for these traits would also become more highly represented in the gene pool. Since reproduction is considered "success," adaptations that lead to better chances of reproduction are referred to as **fitness**. The resulting change in populations over time is commonly called **microevolution**. Darwin summarized this process by calling it "**natural selection**." He argued that over great periods of time and under the necessary conditions, natural selection could result in the development of new species. So a change in a population over time resulting in the formation of a new species would be referred to as **macroevolution.** It often simply goes by the name "Theory of Evolution" for short. Darwin then went on to support his argument by referring to **artificial selection**. An example of artificial selection is the controlled human breeding of wild dogs over hundreds of years that eventually led to many different breeds of domestic dogs. Darwin reasoned that if the human breeding of domesticated animals over hundreds of years could result in the many breeds of dogs, then natural selection over millions of years could result in many different species.

Darwin was still missing one key piece to the puzzle. Although both ancestry and natural selection made sense, one key question remained. Where did the adaptive traits come from that were necessary for populations to change? Darwin was never fully able to answer this question because he did not understand the principles of genetics (he did not have access to the work that Gregor Mendel published just a few years after *On the Origin of Species*). However, he predicted that some sort of change in the then-unknown genetic material could lead to adaptive traits. It turns out that he was right; these changes are called **mutations**. Most of the time when genes are randomly damaged, the result is a loss of gene function leading to either a neutral or negative overall effect on fitness. Rarely does a mutation create a change that provides a trait with a selective advantage; however, it can occur. An example of a mutation leading to a selective advantage is **sickle cell anemia**. This **recessive** disease is due to a mutation in the hemoglobin gene. Only people who inherit two mutated alleles (**homozygous recessive**) have the disease. So this disease is one of many that is the result of a mutation with a negative effect. However, in areas of Africa with high rates of **malaria** infection, something unusual is observed. Malaria parasites infect human red blood cells. This usually leads to a high mortality rate in infants and children. But people who have inherited only one sickle cell allele, which are called **heterozygotes** or **carriers**, are less apt to die from malaria. In addition, they are not fully affected by sickle cell anemia. It seems the people who make some normal and some mutated hemoglobin have a better chance of reproductive success by avoiding both sickle cell anemia and malaria. This, of course, keeps a higher frequency of the sickle cell gene in the gene pool in regions where malaria is common. It does not surprise us that the sickle cell allele is found in much higher frequency in people of African descent than those of European descent.

Now that all of the pieces of the puzzle are together, the process of evolution can take shape. However, it is important to carefully look at some conditions that are helpful for an evolutionary event to take place. First, a small, isolated, selectively breeding population is necessary to facilitate the production of new species by natural selection. A small population is important because within a large population, it is very difficult to drastically change the gene pool. This small population then has to acquire a beneficial mutation that causes its members to have a trait that is environmentally favorable. Over time, this trait must be naturally selected, leading to a change within the population. The Theory of Evolution argues that after many favorable mutations and much time, eventually a new species would evolve.

Small populations can be isolated by a few different events. A **bottleneck** can form when the majority of a population is killed by earthquake, fire, volcano, or some other event, thus leaving a small subset of the original population alive. Likewise, a small population can become isolated when a small portion of the population colonizes a

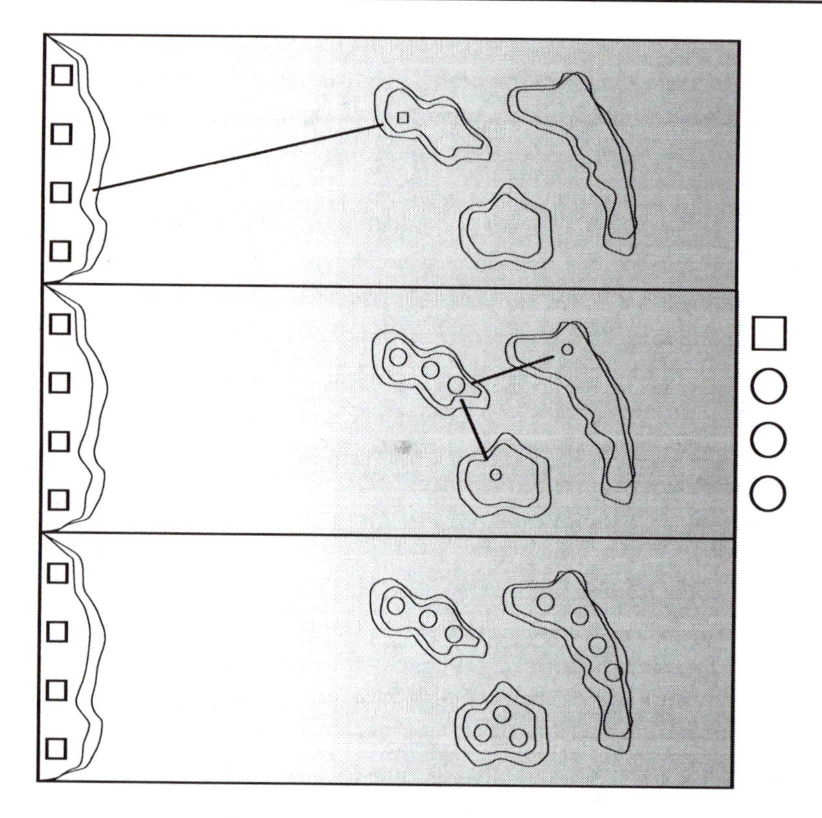

Figure 1. This figure illustrates how Darwin's hypothesis would describe the evolution of 13 finch species on the Galapagos Islands.

new location, leaving the rest of the population behind. Either way, the new small population will probably not have the same gene frequencies that the original population had. This will result in an immediate change in the gene pool called a **founder effect**. This is what Darwin argued had happened on the Galapagos with the different species of finches that he found there. The Galapagos held 13 species of finches that were similar to each other, yet were distinct from those on the mainland. Darwin's idea was that an original small population of finches had flown the 600 miles to the Galapagos with the aide of a tropical storm and colonized one of the islands. Over significant time, and with the aid of many favorable adaptations to the local environmental conditions, a new distinct species evolved. Over time, some of either the new species or the original one then populated a second island, and with time, developed into a third species. As this process was repeated, multiple species evolved (see figure, above).

Darwin used observations that may have gone completely unnoticed to create a theory that could explain how the world came to look like it did. He knew he did not have all of the answers, but a working theory doesn't need to have all of the answers. To illustrate this point, let us take a look at one of Darwin's predictions. Darwin predicted that for his theory to be accurate, the fossil record needed to contain animals that showed gradual change over a long period of time. At the point of his writing *On the Origin of Species*, the fossil record showed distinct jumps only between animals of different types. While scientists have found many more fossils that they would consider transitional, there are still many prominent gaps in the fossil record. Regardless of the gaps, Darwin's theory is still a viable way of explaining the observations that he made.

Questions about this case:

1. What is the simple dictionary definition of "evolve"?
2. In what ways can two populations be isolated from each other besides physically? Describe how cicadas in the Eastern United States are **temporally isolated** from their cousins.
3. Are different breeds of dogs different species?
4. A donkey and a horse mate and the offspring is a mule; yet, a donkey and a horse are considered separate species. How do you explain this?
5. Even in the United States, people of African descent have a significantly higher frequency of the sickle cell allele than those who are not. Would you expect it to stay that way over time? Why?

Questions to go deeper:

1. How do a hypothesis and a theory differ?
2. The words "observation" and "theory" are highlighted in the first paragraph. These are key words that are used to explain the scientific method. What do they mean, and what are the other key parts of the scientific method?

Why I Am Happy a Deadly Disease Runs in My Family

A Case Study of Sickle Cell Anemia, Malaria, and Natural Selection

Malaria infects up to 500 million people per year and kills almost 3 million, a majority of whom are children. According to the World Health Organization (**WHO**) the death toll amounts to about one child every 30 seconds (1). This scourge is caused by a single-cell **protozoan** from the **genus** ***Plasmodium,*** which is transmitted by the bite of a mosquito. Protozoans are **eukaryotic** cells; therefore, infections by such organisms are considered **parasitic**. There are two major species of *Plasmodium* that cause malaria in humans, ***P. vivax*** and ***P. falciparum***. *P. falciparum* infections are much more serious. The first step in a description of a parasite usually involves describing its lifecycle. For humans, the parasite infection begins when an infected mosquito bites the human to take a bloodmeal. Once in the human body, the parasite travels to the **liver**, where it divides **asexually** and is released into the bloodstream. The parasites then infect **red blood cells** (RBCs), where they will either undergo additional asexual divisions or produce **gametocytes**. Either way, the RBCs will rupture and release newly formed parasites that are ready to cause further infection. If another mosquito bites the infected human at this point, it ingests a bloodmeal that contains the *Plasmodium* gametocytes. The gametocytes will then **fertilize** each other and mature over about 10 days as they migrate to the **salivary glands**. This will ensure that the next bite can infect another human host.

In 1955, the WHO declared war on malaria, but by the late 1970s, the organization realized that it had lost. Over that period, the mosquito larvae had shown **resistance** to **DDT**, the insecticide used to reduce the numbers of mosquitoes. Over the same period, the parasites had also shown resistance to drugs that scientists had hoped would severely reduce the number of human infections. Currently, the **distribution** and **prevalence** of malaria looks much like it did 50 years ago, about 20 degrees **longitude** roughly centered on the equator, an area that is sometimes known as the "malaria belt."

With such a grim outlook, what could defeat such a foe? It turns out that a very unlikely and very serious genetic disease can. This presents an interesting story of natural selection. If you look at a map of the distribution of **sickle cell anemia** in Africa, it neatly approximates the similar map of the distribution of malaria (see figure, below). Is this coincidence? Maybe it is just due to the fact that the tropical environment of that area somehow favors both diseases. It turns out that this is not a viable answer because sickle cell anemia is a genetic disease, not an infectious one, and is not affected by the environment as such. In fact, the overlap between malaria and sickle cell anemia is not a coincidence at all. Sickle cell anemia is the result of a single mutation in the **hemoglobin** (Hb) **gene**. Hemoglobin, you will remember, is the molecule in RBCs that carries oxygen to cells. Mutated Hb simply does not function well enough to support a long life; therefore, sickle cell anemia sufferers often die before the age of 30. You are probably thinking, "How can sickle cell anemia help the fight against malaria?" A curious observation was made: people who are sickle cell **carriers** have about a 90% greater chance of surviving to adulthood without contracting malaria than those who have two normal copies of Hb. Remember that humans have two copies of each gene so that a person with one mutated Hb gene will still make enough normal Hb to carry on a relatively normal life. The statistics indicate that malaria kills primarily children, and scientists have found that children who survive malaria can become more resistant as adults. Therefore, as adults they will be more likely to survive further bouts with malaria. This means that being a carrier of sickle cell anemia gives children a better chance to survive malaria, which boosts their overall chance of survival in areas that are **endemic** for malaria, like equatorial Africa.

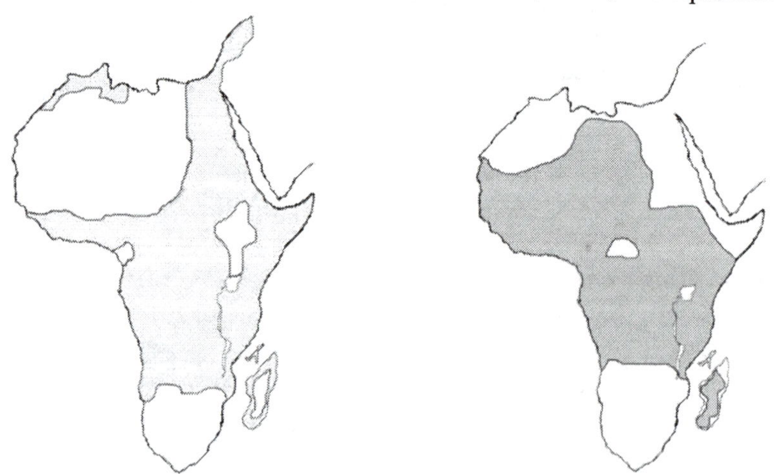

The distribution of sickle cell anemia in Africa (left map) neatly approximates the distribution of malaria (right map) in Africa. Being a carrier of sickle cell anemia gives children a better chance to survive malaria, which boosts their overall chance of survival in areas that are endemic for malaria, like equatorial Africa.

Because of this, sickle cell anemia is quite prevalent in people of African decent. The numbers are startling: 1 in 10 people is a carrier and about 1 in 500 has the disease. In the general population of the United States, 1:150 are carriers, and 1:4,000 have the disease (most of whom are of African descent). The relationship of sickle cell anemia to malaria makes this an interesting study of **natural selection**. Normally, a serious genetic disease will cause those who have the disease to have a lower survival rate. Decreased survival results in fewer individuals reaching maturity and reproducing. In an evolutionary sense, we would say that the disease causes decreased **success**. Success would, therefore, be defined as reproduction. One would expect a disease like sickle cell to cause decreased survival to adulthood and decreased success, resulting in an overall decrease in prevalence of the disease over time. Evolutionary biologists would describe this in terms of the "**gene pool**." For a **population**, the gene pool would consist of all the genetic material that the individuals in the population have. Therefore, the gene pool for a population would have a group of Hb genes, of which some percentage would be mutated. We would expect a genetic disease like sickle cell that causes decreased success to have less of a chance of keeping the mutated genes in the gene pool. Then why is the prevalence of the mutated Hb gene so high in the equatorial African gene pool? In this case, the presence of Hb gene provides a selective advantage to those who carry it. The mutation, while causing a terrible disease, provides a benefit for those living in areas endemic to malaria. By definition this is natural selection: a random natural mutation that results in a better chance at reproductive success. Therefore, we would not expect the mutated Hb gene to provide any advantage to people who do not live in areas ravaged by malaria. This is why the chances of having sickle cell anemia is high only in those areas where malaria is prevalent.

Questions about this case:

1. How might this scenario be different if malaria killed more adults than children?
2. In relative terms, how often do mutations result in a selective advantage?
3. What would be the chance of two carriers for sickle cell anemia having a child who has the disease? What would be the chance they would have a carrier?

Questions to go deeper:

1. How does resistance to anti-malarial drugs come about? How much of a problem is this?
2. Where did the first heritable mutated Hg gene come from?

References:

1. www.who.int
2. www.ascaa.org/comm.htm

How Smokey the Bear Almost Destroyed the Everglades

A Case Study of Ecosystems and Ecology

"Only you can prevent forest fires!" Since 1944 this refrain, uttered by Smokey Bear or, as more commonly known, **Smokey the Bear**, has been one of the most familiar slogans in America. It has been reinforced by one of the most endearing and enduring symbols in Americana, a real live bear who, as a cub, was found clinging to the ruins of a tree after a forest fire in New Mexico. Rescued by fire fighters and eventually sent to the National Zoo in Washington, D.C., this cub became famous as a "spokesbear" for the U.S. Forest Service's fire suppression efforts.

This effort, begun during World War II in response to fears that wildfires might be started by German and Japanese spies, continued for fifty-four years. Posters, television and radio ads, and even comic books were used to enlist every man, woman, and child in the effort to stop forest fires. It was further supported by massive outlays of money, manpower, equipment, and efforts as state and federal agencies did their utmost to stop all fires wherever and whenever they occurred. Their efforts were so successful that the amount of area destroyed by fire dropped from 22 million acres per year to less than 4 million acres a year, with some areas of the nation's forests not burning for decades.

Since this resulted in thousands of acres of beautiful forests being saved, the campaign was in place for decades and, with a few adjustments, continues to be a major effort of the state and federal forest services. These efforts were particularly successful in the Florida Everglades, North America's only **subtropical ecosystem**, located at the tip of the Florida peninsula. From the inception in 1947 of **Everglades National Park** until the late 1950s, all forest fires in the park were vigorously fought. However, ecologists and park rangers began to notice that significant changes were occurring in the unique ecosystems comprising the park. A 1958 study pointed to forest fire suppression as a possible reason for the changes that were being seen and, suddenly, Smokey's slogan was called into question.

Ecologists have long known that ecosystems change over time, a process called **succession**. In **primary succession**, bare rock is slowly transformed into soil, which then provides a home for pioneering species of plants like weeds and grasses. If the climate is suitable, a series of **communities** will move into the area, as each group of plants creates an environment where their own seeds cannot survive. This is due to the fact that some plants require a great deal of sunlight, bare soil, and no competition in order to germinate, grow, and survive. As they fill the area, these conditions change so that their offspring cannot survive. Therefore, new species move in that do not require as much light and can survive without bare soil and with competition. Thus, short grasses that initially populate the area may be replaced by long grasses, which will be replaced by shrubs, and so on. Eventually, a **climax community** will be created that will be composed of plant species that can survive in their parent's shadow, thus replacing themselves, technically, forever.

Today, the majority of succession occurs where the climax community was destroyed to make room for farms and pastures or by forest fires, logging operations, etc. When the farms and pastures are abandoned or the forest starts to recover from the fires or logging, **secondary succession** occurs. This process is virtually identical to primary succession, except that the soil already exists in the area. This process has been particularly well studied in the Southeastern United States, where the original oak-hickory forest climax community was destroyed to clear land for the original farms and plantations of the Old South. Following the Civil War, much of this farmland was abandoned and secondary succession began. The oak-hickory forest climax community is just now returning, 140 years later.

In the Everglades, scientists originally believed that what was present when the park was established was the climax community for that area. They were, therefore, surprised when changes began to occur during the early years of the park. The unique ecosystem known as the Everglades has been termed a "river of grass" because the majority of the area is a shallow and slow-moving mass of freshwater that is slowly heading south toward the tip of Florida. The primary vegetation is composed of **sawgrass**, a vegetation unique to this system. But areas of scrub pines, cypress and **mangrove swamps** and raised areas called **hammocks** also exist throughout the park. However, with the suppression of fires throughout the park, park rangers noticed that changes were occurring in many areas. Species that had been present for years were disappearing and being replaced by other species. Pine and hardwood areas were spreading, sawgrass was disappearing, and the animals that inhabited those areas were also disappearing. Of even more concern, from a human perspective, was the fact that the suppression of fire was allowing large amounts of dead plant material to accumulate. Since forest fires can be prevented only for so long, when these areas did eventually catch fire, the blazes were much worse than normal, due to all the accumulated fuel. The fires were thus much hotter, did much more damage to the ecosystem, and were much harder to put out.

As scientists studied this phenomenon more closely, they began to realize that fire is a natural and important part of many natural communities. What they discovered was that a number of important ecosystems, such as the Everglades, the **long-leaf pine forests** of the Southeast, and the **sequoia forests** of California, are actually maintained by fire. The plant species making up these communities are designed to survive periodic fires, while their competitors, those plants that begin to replace them if fire is suppressed, are not. Since most natural fires, those set by lightning usually, occur regularly, the dead vegetation does not get a chance to build up very much and the fires tend to be smaller and less intense. These types of fire expose the soil, create fertilizer in the ash they leave behind, reduce competition, and, in the **conifer-based** system, even open up the pinecones so the seeds can be released.

With Smokey doing his thing, none of this was happening, and the natural communities, which are called **fire maintained sub-climax communities**, were changing into the climax communities for that particular **climate**. Now, if the fire that maintained these systems was strictly due to humans, this would be a good thing. However, these communities have existed for thousands of years before humans, and fire was a natural part of the system. Therefore, it was our intervention in the natural process that was causing the changes that were being seen in each of these important ecosystems.

In response to this new understanding, park rangers started changing their tactics. Today, many natural forest fires are allowed to run their course if they are not endangering houses, park buildings, and other human structures, or if they are not burning at intensities that are dangerous to the natural ecosystems. Even if natural fires are not occurring, rangers will start **prescribed burns** that are designed to mimic the natural fires, by burning at relatively low intensities. These fires are set only when the weather, moisture content of the fuel, wind, and other factors allow a controlled fire that will not, hopefully, get out of hand.

This new understanding of fire may well have contributed to the recent change in Smokey's slogan. Instead of "Only you can prevent forest fires," Smokey now intones, "Only you can prevent **wildfires**." While this change may have been an attempt to update the slogan, it is also drawing the distinction between forest fires that are set deliberately to help the community or those that occur naturally but are not wild and out of control, and those that are destructive, out of control, and dangerous to the natural and human-made systems. Smokey, thus, may still have his place in protecting these natural systems, but don't be surprised when you are driving through

Yosemite National Park and see smoke but no attempts to fight the fire. In fact, there are often signs posted in parks like Yosemite informing visitors that the proper authorities are aware of a fire but it is not being actively suppressed. It's a new day for Smokey the Bear....

Questions about this case:

1. What does "community" mean in an ecological sense?
2. What is an ecosystem?
3. What are some of the valuable aspects of fire as mentioned in this case?
4. How do we keep forest fires from becoming wildfires?

Questions to go deeper:

1. The case mentioned everglades, long-leafed pine forests, and sequoia forests as three ecosystems in the United States. What are some others?

References:

1. Anonymous. 1996. Fire Management in the Everglades. National Wildlife Federation. www.nwf.org/everglades/fireManagement.cfm
2. Anonymous. 2001. After More Than 50 Years, Smokey Bear has a New Message. Ad Council Forestry.about.com/library/weekley/aa043001a.htm
3. Anonymous. 2006. Fire in Everglades National Park Ecosystems. National Park Service. www.nps.gov/ever/fire/fireeco.htm
4. Anonymous. 2006. History of Smokey Bear. Penn. Dept. of Conservation and Natural Resources. www.denr.state.pa.us/forestry/ffp/history.aspx

Where Have All the Pretty Critters Gone?

A Case Study of Climate, Ecosystems, and Diversity

It was 1984, and I was teaching my second college-level ecology class in Southern California. One beautiful Saturday during the Fall semester, I took a busload of students north from the Los Angeles area to spend the day looking at a variety of **ecosystems,** including rocky seashores, sandy seashores, and tidal pools. It was the same trip I had taken during my first such class in 1982. When we got to the tide pool area, I gathered the students together and explained what they should be looking for. I told them about the sea urchins, sea anemones, ochre sea stars, baby octopi, and other cool things they should see, based upon what we had seen two years ago. After my introduction, the students fanned out looking for any and all creatures they could find, excited to see who could find the best critters.

Pretty soon, however, the students started coming back, disappointment written on their faces. "There's not much here, Dr. K. Are you sure we are at the right place?" As I began looking around myself, I, too, began to wonder if I had mistakenly taken the students to the wrong beach. It sure looked the same in terms of the landmarks I remembered. It was the same time of year and the tide was about the same level it had been last time, but something was definitely wrong. Where once there had been tide pools full of abundant marine life, there were now tide pools with few organisms. The ones that were there were present in good numbers, but the **diversity**, the variety, and the numbers of each type present in the area had been greatly reduced.

Unfortunately, I had no explanation as to why things had changed so drastically. In fact, it wasn't just dumb old me who didn't know what was going on ,as science had yet to figure out exactly why such drastic changes occurred periodically throughout much of the world. We knew that droughts, flooding, severe tornado, and hurricane seasons and other significantly unusual weather events happened every once in awhile and, of course, we had a pretty good handle on what caused the world's **cli-**

mate, which can be defined as the long-term atmospheric conditions over a large region of the earth. However, we had no good explanation for why the unusual events periodically occurred.

Typically, the earth's climate has been pretty easily explained and, within a person's lifetime, pretty consistent. We know that because the earth is a sphere that is tilted at about a 12° angle and which revolves around the sun, different parts of the earth get hotter than other parts. Since the **equator** is the closest part to the sun, it gets hotter. This causes the air to heat up and rise. This creates a **low pressure** area. As the air rises, it cools off and loses its ability to hold water. This causes it to rain in these low pressure areas. The lowest section of the Earth's atmosphere is the **troposphere**. Once the rising air hits the upper portion of the troposphere, it can only move north and south. Eventually, this air gets colder and colder, which causes it to be heavier than the surrounding air. This causes it to start falling toward the Earth, creating **high pressure** areas. Now the falling air starts to heat up and it gains the ability to hold more water, creating dry areas. This process occurs around the 30° north and south **latitudes**.

Now that the falling air has hit the Earth, it has to go north and south. As the air moves across the Earth's surface, it heats up and, around 60° north and south latitudes, it rises, creating another low pressure and rainy area. The process repeats itself at the north and south poles, where cold air falls and another dry, high pressure area develops.

These rising and falling areas of air cause the rainy and dry regions of the Earth. The large-scale wind patterns also result from this up and down movement of air. This is because once the air comes down, it has to move north and south across the surface of the Earth. The air moving down eventually replaces the air moving up and, voila, you have wind! However, to further complicate things, instead of moving straight along the Earth's surface, the wind veers either to the right or to the left due to the **Coriolis Effect**. This is what causes the veering of anything moving across the Earth's surface due to the rotation of the Earth on its axis. Due to this, you have things moving to the right in the Northern Hemisphere and to the left in the Southern Hemisphere.

This means that, normally, the air is moving from South America toward Australia. This movement is enhanced by the fact that, in most years, there is a localized low pressure area over Australia and a localized high pressure area over South America. All of this creates a flow of surface winds that moves west from South America to Australia. As the wind moves across the water, it also starts the movement of surface water in a westerly direction. In order to replace the surface water moving westerly, the bottom ocean water moves from west to east. This means that cold bottom water is moving up in the area near Chile. This **upwelling** water carries with it great amounts of nutrients, which serve as fertilizer for the **phytoplankton**, the microscopic **photosynthesizers** that are at the bottom of the food chain. These organisms at the bottom of the food chain are often called **producers**. Because of this movement of cold water, the oceans off the west coast of South America are incredibly productive.

So, what does this all have to do with the critters found in the tide pools of Southern California? By definition, **tide pools** are those small bodies of water that are left exposed by the falling tides, which occur twice a day, and are then covered by the rising tides, which also occur twice a day. Due to the exposure that they face twice a day, tide pool organisms have to be uniquely adapted to daily, or **diurnal**, changes. Ecologists would say that organisms that live in these pools fill a specific **niche**. Because they are isolated from the main ocean several times a day and because the rising and falling tides involve a lot of wave action, these organisms have to be able to withstand daily changes in temperature, salinity, and the physical violence of the wave action. These organisms are, thus, amazing creatures.

However, while they are able to withstand these daily changes, they cannot tolerate long-term changes in the temperature of their water. When normal conditions exist in the area between Australia and South America, the water along California is pretty cold. This is because the global wind patterns also create global water patterns, called **currents**. The water moving along the coasts of Southern California is coming from the north, carrying cold water rich in oxygen and nutrients. This provides an ideal environment for the types of tide pool organisms that were there the first time I took my students to the tide pools.

Unfortunately, in 1982-83 a severe **El Niño** pattern developed that significantly altered the air and water flow in the Pacific Ocean, causing portions of its waters to increase up to 6°C in temperature. These El Niño patterns occur periodically whenever the air pressure in the middle of the ocean between Australia and South America lowers, creating an abnormal low pressure area. This causes the normal low pressure area over Australia to be replaced by a weak high pressure area. The strong **trade winds** that normally flow from east to west become much weaker, as does the movement of surface water from east to west. Instead of cold bottom ocean water moving up toward the coast of South America, warm surface water remains there.

These changes mean that both local and global changes occur. Depending upon the severity of the El Niño episode, the United States and South America may experience very wet weather, while parts of Australia, Indonesia, and Africa may experience droughts. As the water warms up, oxygen content drops, and as upwelling of bottom water decreases, few bottom nutrients are available. These changes are devastating to ocean creatures, starting at the bottom of the food chain. The drastic changes brought on by the 1982-83 El Niño reduced the amount of fish caught off the coast of South America by 50%. While the changes that occurred further north, along the coast of Southern California, were not nearly as drastic, the effect on coastal organisms was evident in the reduced diversity we were seeing in the tide pools.

These changes were seen up and down the coast of California, as Southern California started to look more like Baja California and Northern California changed to look more like Southern California used to. Just a one to two degree change in water temperature can cause organisms to disappear from an area, which is exactly what we were seeing. Fortunately, these changes are not long lasting and, within a year or two, the conditions become "normal" and, slowly, organisms return to their old areas. On the downside, however, these El Niño periods are steadily increasing in frequency. Historical data indicates that they used to occur every 100 or 150 years. They slowly but steadily began occurring every 40 or 50 years, and now, in the last several decades, have begun to appear every 10 to 15 years. In addition, they also seem to be lasting longer and longer.

Almost two decades after my first experience with El Niño, I again saw evidence of the impact such changes can incur. This time it was not the small invertebrates living in the tide pools that told the tale. It was the large number of dead seals and sea lions my class observed washed up along the shore as we traveled around Southern California in the late 1990s. These organisms illustrated the impact of El Niño on the larger mammals. Due to **predator-prey relationships**, the large mammals that are at the top of the food chain and sometimes called **tertiary consumers** depend upon the large fish (**secondary consumers**). The large fish feed on the small fish (**primary consumers)** that feed on the phytoplankton (our producers) that were being killed by the warm water temperatures. During the season, the news was constantly reporting on weakened seal and sea lion pups being rescued along the coast, reports supported by video of emaciated animals being aided. It was vivid evidence of the devastating effects such seemingly inconsequential changes in one part of the world can have thousands of miles away.

Question about this case:

1. In the talk about the food chain, we left out the decomposers. What are they and why are they so important to the food chain?
2. The "food chain" is a bit of an old school term; it is probably better called a "food web." Why is this?
3. The case mentions the concept of a niche. Define what a niche is.

Questions to go deeper:

1. Our tides rise and fall approximately twice a day. What causes tidal changes?
2. If things veer different directions in the Northern and Southern Hemispheres due to the Coriolis Effect, how does this change the way toilet bowls flush or hurricanes and monsoons twist?
3. Where did the trade winds get their name?

References:

1. Dybas, C. 1995. Dial-a-Scientist Studying El Niño. NOAA. www.pmel.noaa.gov/tao/elnino/dial-a-scientist.html
2. Kay, J. 2006. A Warming World: the Difference a Degree Makes in Seashore Sea Change. San Francisco Chronicle. www.sfgate.com/cgi-bin/article.cgi?file=c/a/2006/01/16/ MNG9UGO2DO1.DTL
3. Pidwirny, M. 2006. Introduction to the Atmosphere: El Niño, La Niña, and the Southern Oscillation. Univ. of British Columbia.
4. Wikner, B. 1998. Tide pool functionality: the mechanics and the variables. www.pleasanton.k12.ca.us/avhsweb/thiel/creek/ap98/Wikner/rea.htm

To Biopsy or Not to Biopsy

A Case Study of Bone, Orthodontics, and Osteoporosis

During the second year of medical school, the physiology and pathology of each organ system was taught in great detail. As Tim took his seat, he was excited to learn about the pathology of connective tissue and bone. As the room darkened, the professor projected the first image on the screen. The photograph was of a man with a noticeable bulge in the center of his chest. The professor asked, "Assuming that you have taken a complete medical history, what would be the first thing that you would do to begin to diagnose this lump?" To an overeager medical student, a lump can only mean one thing: cancer. A student near the front raised his hand and said, "I would take a needle biopsy." To which the professor replied, "I am sorry, but you have just killed your patient!" The professor had just set a trap, and the student had fallen right into it. The professor continued, "This is an aortic **aneurism** that has pressed against the **sternum.** In response to the pressure, the bone cells have dissolved the bone…". What Tim was seeing on the screen was the largest blood vessel in the body, the **aorta**, bulging through the open sternum. Had the medical student actually stuck a needle into the lump to get a biopsy, the aorta would have burst, and the patient would have bled to death in seconds. The professor had laid the trap perfectly in order to teach this important concept to the medical students: it is much easier and safer to **palpate** a lump before you proceed. However, since this lecture was on bone and connective tissue, there was another essential question. How could the pressure of the aneurism cause such a massive removal of bone?

Bones are alive; they contain living cells. Even the densest parts of bone, called **compact bone**, have microscopic channels where bone cells live. In addition, there are other channels in compact bone that supply these cells with oxygenated blood. These cells have many specific functions that help to maintain healthy bones. Bones are not static; they can change under certain conditions (such as pressure). Two specific types of bone cells, known as **osteoclasts** and **osteoblasts**, are responsible for the bones' ability to change. Osteoclasts work to remove bone, while osteoblasts work to add bone. Both of these cell types respond to a variety of stimuli to help regulate when and where bone mass is added or removed. **Calcium** is the key mineral in this process, since it is the mineral that makes bones hard. Essentially, in order to add or remove bone, these cells add or remove calcium. In the case of the patient with the aneurism, pressure provided the stimulus to which the osteoclasts responded. When pressure is applied to bone, osteoclasts begin to remove calcium from the bone. Over time, this can lead to a complete loss of the portion of bone that is receiving the stimulus.

The use of modern orthodontics is another good example of how osteoclasts and osteoblasts can regulate the movement of bone. To straighten or move teeth, a small amount of pressure is placed on the tooth by a series of wires or rubber bands. If you have had (or currently have) braces, you realize that this does not seem like a small amount of pressure. But, when the pressure causes the tooth to push against the bone, the osteoclasts begin to remove calcium, and the tooth begins to move. This process can have noticeable effects (loosening of teeth) in a matter of a few days. Osteoblasts then begin to deposit calcium where there is an absence of pressure. Therefore, as the tooth is pulled by the wires, the osteoclast removes calcium so that the tooth can move in the desired direction. The osteoblasts then deposit calcium into space where the tooth was. This is often called **bone remodeling**. In fact, it takes an osteoblast much longer to deposit enough calcium to fully harden bone than it takes for an osteoclast to remove the calcium. For this reason, many people who have had their braces removed are given a retainer to wear for a year. This is done to keep their teeth straight until the osteoblasts have sufficient time to finish their work.

As people develop and mature, we find that their bones develop and mature, as well. In fact, osteoclasts and osteoblasts respond to the same hormones that tell other parts of the body to grow, to develop, and to mature. People reach a peak **bone mass** around age 30. From that point, osteoclasts begin to remove bone slightly faster than osteoblasts can replace it. Therefore, people tend to loose about 1% of their bone mass per year. When women reach **menopause,** they produce dramatically less **estrogen,** which can accelerate bone loss. The problem with bone loss stems from the fact that **weight-bearing** bones are not solid. The previously mentioned compact bone makes up an outer layer that surrounds a core of **spongy bone** and a large cavity for the **bone marrow.** If calcium is lost from the compact bone to the point where the bone becomes weakened, the spongy bone and marrow cannot assist in supporting the necessary weight. Ultimately, bone loss can become so severe that bones become brittle. This disease is called **osteoporosis**. In fact, 80% of the people who suffer from osteoporosis are women (1). Sadly, a woman has about a 50% chance of having an osteoporosis-related fracture sometime in her lifetime (1). Because of the prevalence and severity of osteoporosis, physicians sometimes prescribe **hormone replacement**

therapy for women who have significant bone loss during menopause. This treatment simply provides extra estrogen to signal the bone cells, so that they will add more calcium to the bone.

So, what are some good practices to help to keep bones healthy? Since bone loss can be an issue for both men and women, it is essential to eat a diet high in calcium. In addition, physicians suggest a diet that is also high in **vitamin D**. This is because vitamin D helps your body absorb the calcium that is ingested. Physicians also suggest regular exercise. Osteoblast activity increases when a workload is placed on bones. Even an exercise as simple as vigorous walking can aid in gaining or maintaining bone mass. People should consult a professional in order to determine an appropriate exercise regimen. Scientists have also shown that smoking and alcohol consumption can accelerate loss of bone mass. Therefore, eliminating these habits will also help to prevent osteoporosis.

Questions about this case:

1. What are some of the other stimuli that bone cells respond to besides pressure?
2. How could the concept of bone remodeling be used in repairing bone fractures?
3. What are some food items that are rich in calcium?
4. What are some food items that are rich in vitamin D?
5. What is a non-dietary source of vitamin D?

Questions to go deeper:

1. What is the function of the bone marrow?
2. Why is it an especially good idea to enrich milk with vitamin D?

Reference:

1. http://www.nof.org/osteoporosis/diseasefacts.htm

Paul and Liz's New Oven

A Case Study of Carbon Monoxide Poisoning

Their new house didn't come with appliances, and as far as Paul and Liz were concerned, that was just fine; a 1930s vintage home deserved a vintage stove. It did not take much time searching the newspapers to find a vintage gas O'Keefe and Merritt stove that would fit great. It was even in their price range! They picked it up on a Sunday afternoon and cooked with it that night. After dinner, both Paul and Liz did not feel very well, so they watched a little TV, went upstairs, and went to bed early. They woke up feeling great. Over their typical breakfast of coffee and toast, they planned to cook a great dinner together on their new stove. They really wanted to see how the oven worked. They hurried off to work with plans to meet back at home by 5:30 with everything necessary to make dinner, including baking bread in their "new" old oven.

Dinner went off without a hitch. But, by the time the bread had risen and the oven was preheated, they were both feeling ill again. This time it was worse. Their headaches were mild at first, but they grew worse as time went by. They both were nauseous by the time the bread was finished, and they did not even feel like trying it. They couldn't get to bed fast enough. Like the day before, they woke up feeling great. They had no idea what was happening to them.

All fuel-burning appliances, including Paul and Liz's oven, produce **carbon monoxide** (CO) as a result of incomplete combustion. In most cases, appliances that are in good working order produce only very small amounts. But, Paul and Liz's old O'Keefe and Merritt wasn't in good working order, as they had thought. In fact, the burners for both the oven and stovetop were in such poor shape that significant amounts of CO were being produced.

Why is this a problem? The answer is found in understanding how the circulatory system carries **oxygen** and **carbon dioxide** (CO_2) around the body. Blood gets pumped back to the heart from all over the body in **veins**. At this point, the **red blood cells** (RBCs) are carrying very little oxygen. The heart then pumps the deoxygenated blood (full of CO_2) to the lungs so that the RBCs can give up the CO_2 that they were carrying and pick up oxygen. The key molecule in the process is **hemoglobin**. RBCs are packed full of hemoglobin. This molecule has only two jobs: to carry oxygen from the lungs to cells throughout the body using vessels called **arteries** and to pick up CO_2 from the cells and carry it back to the lungs.

Blood normally gets pumped to the lungs through the **pulmonary arteries**. Once in the lungs, these arteries branch into very small vessels called **capillaries**. The capillaries make a vast network that tightly surrounds the air sacks in the lungs, called **alveoli**. When air is breathed in, gas molecules, such as oxygen, **diffuse** from the alveoli into the capillaries. The distance between the capillaries and the alveoli is very short; therefore, the gasses do not have to travel very far during diffusion. The hemoglobin in the RBC then picks up the diffused oxygen, and releases the bound CO_2. This CO_2 then diffuses out into the alveoli of the lungs and is expelled with the air that is breathed out. In the case of Paul and Liz's illness, the blood that reached the lung encountered a mixture of oxygen and CO that had been breathed in. These gases then diffused from the alveoli into the RBCs. Hemoglobin still gave up the CO_2 that it was carrying, but it prefered to pick up CO rather than oxygen. Normally, if a person breathes in small amounts of CO, it will diffuse into the RBCs and bind to hemoglobin. When the levels are low, there is plenty of hemoglobin left to bind to oxygen. In such a case, there would be few, if any symptoms. However, if a person breathes in air with high amounts of CO, as Paul and Liz did, the hemoglobin in the blood binds to CO before it binds to oxygen. This means that the blood carries little oxygen, thus leaving the cells of the body starved for O_2. Prolonged exposure to CO can lead to cell death. The brain is very sensitive to the lack of oxygen, so a headache is usually the first symptom. If the body encounters persistently high levels of CO, then death from lack of oxygen is imminent.

The critical need for oxygen can be illustrated by how carbohydrates are broken down to produce **adenosine triphosphate (ATP)**. ATP stores chemical energy for the cell. A **carbohydrate** like **glucose** (which has 6 carbon atoms) can be broken down into 2 **pyruvate** molecules (with 3 carbons each) in a process called **glycolysis**. If oxygen is not present, the pyruvate is then turned to **lactic acid**, and only a small amount of ATP is produced (not enough for human survival). This is called **anaerobic respiration** (see the equation on the following page). However, if oxygen is present, **aerobic respiration** can proceed. In this case, pyruvate can be further broken down using the **Kreb's Cycle,** which produces another small amount of ATP, six CO_2 (which you breathe out), and what are called **activated carrier molecules**. These activated carriers are activated when they are given electrons during glycolysis and the Kreb's Cycle.

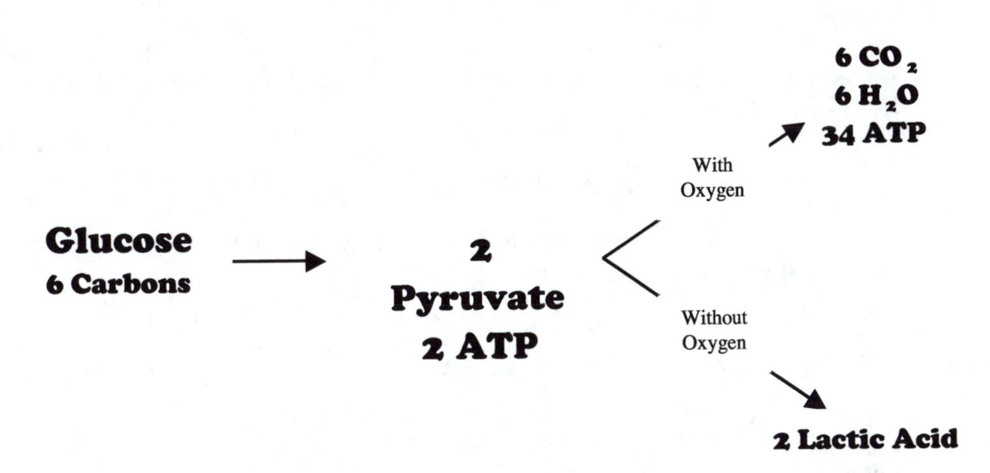

The carriers then donate those electrons to molecules in the **electron transport chain**, which sequentially pass the electrons along to the next members in the chain. This process of passing of electrons releases enough energy to make a large quantity of ATP. The final member in this chain then passes the electron to oxygen to carry it away. Without oxygen, the electron transport chain gets backed up; therefore, there is no place for the members of the chain to pass their electrons. Without the electron transport chain, the large quantities of ATP necessary for survival cannot be made. Consequently, without oxygen, cells would die from a lack of ATP.

The Center for Disease Control (CDC) reports that more than 500 people die every year from accidental CO poisoning. Fortunately, Paul and Liz did not. Each time that they would cook, the CO levels would rise to dangerous levels. But as soon as they were done, they would turn the oven off. This caused the CO production to stop. Had the problem been with their furnace or water heater (something that would have remained on throughout the night) they would have probably died the first night. Some astute neighbors heard about what had happened and loaned them a CO detector. They plugged it in next to the stove and turned the stove on. The detector showed a dangerous level of CO was produced very quickly. This is why the CDC suggests that fuel-burning appliances be correctly installed and inspected yearly. They also suggest purchasing CO detectors; they are relatively inexpensive and save lives. For Paul and Liz, the repair costs for their beloved O'Keefe and Merritt were more than the initial cost; it was an expensive lesson.

Questions about this case:

1. How would this all be different if carbon monoxide bound equally as well to hemoglobin as oxygen did?
2. Why do muscles "burn" when people exercise vigorously for an extended period of time?
3. What is the best immediate treatment for carbon monoxide poisoning?
4. Where in the cell is glucose broken down? Where does the Kreb's Cycle and electron transport chain occur?

Questions to go deeper:

1. Yeast can survive by breaking down glucose in the absence of oxygen. How does this work? What are the by-products?
2. Emphysema and pneumonia both prevent oxygen from binding hemoglobin. How are these different from CO poisoning?

Reference:

1. http://www.cdc.gov/nceh/airpollution/carbonmonoxide/cofaq.htm

The Best Hamburger Money Can Buy!

A Case Study of Fast-food and Nutrition

Where can you get the best hamburger money can buy? When this question is posed to any of my friends, the answer without doubt will be In-N-Out Burger®. My favorite meal at In-N-Out Burger is a Double-Double® with fries and a vanilla shake. I could probably eat this every night for dinner and never get sick of it! While you might debate this choice, one thing is sure: our choices of fast-food meals have become tremendously important. In fact, fast-food in general has become a central component of the American diet. Every year, Americans spend more money on fast-food than they do on higher education (1). On average, Americans spend more than 40% of their food money budget on food eaten away from the home (1).

After reading this you may ask, "What is the big deal?" The answer is **obesity**. It turns out that as the amount of fast-food consumption has been on the rise over recent years, so has the incidence of **obesity**. Over the past 10 years, the percentage of Americans that are obese has nearly doubled, regardless of age and gender. Obesity is defined using the **Body Mass Index**, or BMI. BMI is defined as your weight in pounds divided by height in inches squared, times 703. For example, a person who is 5'6" tall and weighs 185 pounds would have a BMI that would be calculated as $(185/66^2) \times 703 = 29.9$. For adults, obesity is defined as a BMI of 30 or higher and severe obesity would be a BMI of 40 or higher. The problem with obesity is that it can be directly linked to a large number of debilitating and possibly fatal diseases, such as **diabetes, hypertension,** and even **cancer**. So, if I did eat this favorite fast-food meal for dinner every night, I might not get sick immediately, but I would undoubtedly gain weight (unless I radically altered the rest of my life). This is because this single meal (Double-Double® with fries and a shake) contains almost a day's worth of calories, according to the **U.S. Department of Agriculture (USDA) Dietary Guidelines for Americans** (2). The USDA provides these guidelines to help people eat healthy and balanced diets. The guidelines are based upon the principle that a low-fat, low-calorie diet with regular exercise is the best formula for long-term weight loss and maintenance. This does not mean the USDA rejects the value of diets like the Atkins Diet, but it believes that these types of diets are generally short-term solutions. Therefore, the USDA guidelines are based on a diet that consists of 2,000 calories a day. In addition to the specified caloric intake, the USDA also recommends that less than 30% of those 2,000 calories comes from **fat** and less than one-third of that (or 10% of the total) comes from **saturated fat**. Furthermore, the guidelines suggest an intake of less than 300 mg of **cholesterol** and 2500 mg **sodium** per day, as well.

The reason that the USDA suggests people limit their intake of fat, saturated fat, sodium, and cholesterol is that excess intake of each of these is linked to significant disease. High-fat diets are typically high-calorie diets that promote weight gain and obesity. In addition, diets high in saturated fat and cholesterol are linked to high blood cholesterol and heart disease. High-sodium diets have been shown to be a factor in hypertension. With this in mind, let's take a look at my favorite meal, and see how it compares to the suggested guidelines. The total meal (Double-Double® with fries and a shake) contains 1,750 calories, 96 grams total fat (of which, 48 are from saturated fat), 210 mg cholesterol, and 2035 mg sodium (3). To calculate the percent calories from fat, simply multiply the number of fat grams by 9 (calories/gram) and then divide by the total calories from the meal. Therefore, this meal gets 49% of its calories from fat (96 x 9/1750 = .49). Likewise, this meal gets 24.6% of its calories from saturated fat (48 x 9/1750 = 0.246). Obviously, these values far surpass the guidelines! To determine how the calorie, sodium, and cholesterol content on this meal compares to the guidelines, simply divide those values by the guideline value to generate the percentage of the daily suggested intake. When you divide 1,750 calories by a 2,000-calorie diet, you get a meal that contains 87.5% of the suggested daily intake! For sodium and cholesterol, it would calculate to be 85% and 70% of the suggested daily intake, respectively! All this in one meal!

The USDA also provides guidelines for other essential dietary components, such as **vitamins, minerals, and fiber**. Most of these nutrients are expressed in nutritional information as a percentage of the daily intake (based upon the standard 2,000-calorie-a-day diet). For those that are not expressed as a percentage, the calculations are just like the ones above. For example, the USDA suggests the intake of at least 25 grams of fiber per day. If your breakfast cereal has 4 grams per serving, then it contains 16% (4/25 = 0.16) of the minimum daily suggested intake for fiber. In a Double-Double® with fries and a shake, there is a total of 5 grams of fiber (20% of the suggested intake). This means that if I want to keep with the guidelines for the day, I would need to eat 20 grams of fiber in the remaining 250 daily calories (2000 – 1750 = 250 calories)! Wow!

Therefore, should we conclude that if people want to be healthy, they should never eat fast-food? The answer is no for two reasons. First, these guidelines are intended to be simply that, guidelines. A person could conceivably eat a high-fat meal on occasion and balance it with other more healthy foods and over a longer period of time, and stay within the guidelines. Second, if you are careful to read the nutritional information (which is available for almost everything you eat), you can find some fast-food meals that fit the guidelines much better than my personal favorite. A more appropriate conclusion could be that people should eat unhealthy foods in moderation and read the nutritional information carefully so good food choices can be made.

Questions about this case:

1. BMI takes into account only height and weight. What might be some of the problems with such a simple calculation?
2. Should the 2,000-calorie daily intake apply to everyone? Why or why not?
3. There is scientific evidence that some fats are actually good for you (in moderation). Which ones? Why?
4. Find the nutritional information for your favorite fast-food meal and compare it to the guidelines for calories, fat, saturated fat, sodium, and cholesterol. How does it compare? If you don't have a favorite, choose one as an example.
5. What types of food are high in fat, saturated fat, sodium, and cholesterol?

Questions to go deeper:

1. What is the food pyramid? How do the USDA guidelines help to explain its usefulness?
2. What are trans fats? What is the scientific community currently saying about these?

References:

1. Schlosser, E. Fast Food Nation: The Dark Side of the All American Meal. HarperCollins, 2001
2. www.usda.gov
3. www.in-n-out.com/nutritional_info.asp

The Amazing Shrinking Physician

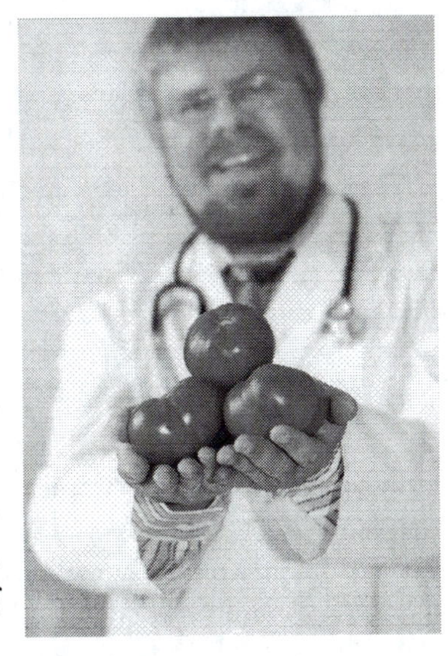

A Case Study of Diet and Nutrition

"As a physician, I kept having to tell my patients to do what I say, not what I do. I just got tired of being a hypocrite," Nick told me once.

Nick was severely obese. He knew that as a physician, he needed to counsel his patients to eat right and exercise in order to maintain a healthy weight. How could he advise his patients to do that if he would not even do it himself? Nick did not make excuses; he knew he had a problem. But, Nick was busy. He often had lunch and dinner meetings with important people who would buy his meal, so that they could get some of his time. "When I did not have meetings, I was so busy seeing patients that I ate out of candy machines," he would say with some embarrassment.

When Nick finally decided to do something about his problem, he knew only a radical solution would work. Nick decided that a year of a nutritionally balanced liquid diet would be required. Can you imagine going on a one-year diet and having nothing but liquids? It was nearly inconceivable, especially for someone so dependent on food. Therefore, Nick came up with a great idea. He decided to quit his job for a year and travel to all 50 states. His trip included seeing a baseball game at every major league stadium. This was his way of providing a diversion. In his mind, desperate times called for desperate measures. "I did not even know how much I weighed," he would say later. "My bathroom scale went up to 350 pounds and since I maxed it out, I figured I was just over that." When Nick finally got on a scale at the beginning of the journey, he weighed an amazing 467 pounds. He had no idea that he was closer to 500 pounds than to 300 pounds! Nick's goal was to weigh 200 pounds, and his motto was "to become half the man I used to be."

Over the following year, Nick accomplished his goals. He remained faithful to his diet, and to the required weekly doctor visits to monitor his progress and health (extreme weigh loss can be very stressful on the body). He also traveled to all 50 states and saw baseball games in every major league stadium. Nick returned home to a hero's welcome and weighed 197 pounds at the time of his first solid meal. "I don't consider myself successful yet," Nick would say later. "Over 90% of people who lose 100 pounds or more gain it back, so talk to me in a couple of years from now about success."

Over the next couple of years, Nick was able to keep the weight off, but it required a completely new way of looking at food and eating. Without the confinement of the liquid diet, Nick had to figure out how to eat to maintain his weight for the rest of his life. The **United States Department of Agriculture** (USDA) has dealt with this type of issue for many years and has created the **food pyramid** to help people understand how to eat a healthy diet. Their dietary recommendations are based upon a diet that supplies a wide variety of **vitamins and minerals**, while being low in **fat** (especially **saturated fat**) and **simple sugars**. The basic idea is that if you eat according to the pyramid, you will do a fairly good job of eating a complete, well-balanced diet without having to be a nutritionist. Interestingly, the current pyramid is actually 12 different pyramids, which are tailored to the age, sex, and activity level of a person (1). What they all have in common is that they suggest that the largest portion of one's diet should be **complex carbohydrates,** followed by vegetables (especially green and yellow ones), fruits, dairy products, proteins (lean meats, nuts, and beans), and fats, respectively. Each pyramid has an optimum calorie intake and the number of servings for each level will vary between the different pyramids. For example, a 20-year-old female who exercises moderately should eat about 2,200 calories per day in the form of 7 servings of grains, 3 cups of vegetables, 2 cups of fruits, 3 cups of dairy, and 2 servings of lean meat.

Along with the pyramids, the USDA has published some guidelines to help people make healthy choices when it comes to their diet (2). They suggest limiting total fat intake to less than 30% of total calories, and saturated fat intake to less than 10% of total calories ingested. The USDA also suggests limiting the intake of **trans fats** as much as possible. This sounds confusing, but it is not. If you read the nutritional information label on a food item, it will have a listing of fat and saturated fat content for that item. Often, the food manufacturer will do the calculations for you, and will provide the percentage of calories that come from total and saturated fat. If the packaging does not provide that percentage, then it is very easy to calculate. For example, let's say that you are eating your favorite ice cream and you find on the label that in a half-cup serving there are 210 calories, and 12 total grams of fat (of which 6 grams are saturated fat). All you need to know is that there are 9 calories in each gram of fat. Therefore, there are 108 calories from total fat (12 x 9 = 108) and 54 calories from saturated fat (6 x 9 = 54). To get the percentage of fat (or anything else), just divide the fat calories by the total calories: 108/210 = 51% calories from fat and 54/210 =

Food Type	Serving Size
Complex Carbohydrates	1 slice of bread 1 cup ready-to-eat cereal 1/2 cup cooked cereal, rice, or pasta
Vegetables	1 cup raw, cooked, or juice 2 cups raw greens
Fruit	1 cup raw cut or juice 1 medium piece 1/2 cup dried
Dairy	1 cup milk or yogurt 1 1/2 ounce cheese
Protein	3 ounces lean meat 1 egg 1 ounce nuts or 2 tbsp peanut butter 1/2 cup cooked beans

26% calories from saturated fat. As you would have probably expected, ice cream is a high-fat food. In addition to fat recommendations, the USDA also suggests that people limit their intake of **cholesterol** to 300 mg per day and their intake of **sodium** to 2,400 mg per day. These numbers don't require calculations. Most nutritional labels will tell you how many mg of each of these is present and what percent of your recommended limit it represents.

Part of the problem with Americans is the size of the portions that we eat. The USDA expresses the serving suggestions in terms of ounces. So, back to our 20-year-old female who exercises moderately, and should eat 7 servings of grains a day. How much is that really? A **serving** of grain is defined by the USDA as a slice of bread, a cup of ready-to-eat cereal, or a half-cup of cooked cereal, rice, or pasta. This sounds fine, but do you really only eat one cup of cereal for breakfast? The guideline is for a cup, not for how much you can fit in the bowl. The point is not that cereal is bad for you, but that the guidelines are designed so that you can eat within an energy limit (2,200 calories for our 20-year-old female) and only consume the necessary nutrients. Look at the table above for some information on other serving sizes. Notice that a serving of meat is 3 ounces, which is only about the size of a deck of cards! A quarter pound of meat is more than one serving (4 ounces). The point here is that how much you eat can be just as important as what you eat. Just take the ice cream example from the previous paragraph. One serving is only half a cup. How much ice cream do you normally eat? For reference, Ben and Jerry's ice cream is sold in pint containers, which hold 4 servings.

Of course, not all dietary recommendations have to do with what is not good for you; many have to do with compounds that are necessary for you to survive. For **essential vitamins and minerals**, food labels usually express values as a percentage of a daily suggested amount. Vitamins are broken into two groups: **water soluble** and **fat soluble**. The fat soluble vitamins are **A**, **D**, **E**, and **K**, and the water soluble ones are **niacin**, **folate**, **C**, and the **B** vitamins. Minerals are atoms that are key components for proper cellular function. These are often metal atoms that are required for other molecules to function properly. Examples of these minerals are **iron, calcium, zinc, magnesium, iodine,** and **copper**. As you look at the following table with examples of food groups that are good sources of these vitamins and minerals, it becomes clear that eating according to the food pyramid should supply each of these compounds in sufficient quantities. Over the years, some food manufacturers have supplemented foods with additional nutrients. These foods are called **fortified** or **enriched**. Many foods are enriched, but the best example of an enriched food is milk. Almost all milk processors add vitamin D to their milk to ensure that children get sufficient amounts of this vitamin that is essential for bone development. A great example of mineral fortification is the addition of iodine to table salt. **Iodine** deficiencies can result in the development of a **goiter.** Therefore, adding iodine to salt almost guarantees that everyone has enough iodine in their diet.

The previous paragraph mentioned essential vitamins and minerals, but we neglected to point out that this simply means that the body cannot produce them on its own. As a result, they must be consumed in the diet. There is a similar concept with **amino acids**. Amino acids are linked end to end to make **proteins**; therefore, when you eat foods containing protein, your body will break apart the protein and reuse the amino acids to make proteins for your cells to use. There are 20 different amino acids, of which 8 are essential. This means that our bodies can make 12 of the amino acids, but 8 of them we must get in our diets. A protein source that contains all 8 of these essential amino acids is said to be **complete**. Meats and diary products are complete protein sources, so for people who do not eat these foods, finding a complete protein source can be more challenging. Tofu and nuts, for example, are <u>not</u> complete protein sources. So for a vegetarian or vegan diet, carefully choosing a variety of protein sources makes ingesting all 8 essential amino acids relatively difficult.

What about physical activity? The USDA suggests that for long-term weight loss or maintenance, a complete diet should include regular exercise. Ultimately, we must balance calorie intake with calorie use. Furthermore, exercise does not necessarily mean strenuous workouts. Some people may not need strenuous exercise. For many, simply walking 30 minutes a day is a great first step (no pun intended). Consulting a physician may be a necessary step for designing an appropriate exercise plan for you.

Nick figured this out for himself and put his thoughts down in a book called *My Big Fat Greek Diet*. He has some great thoughts on learning how to and why to eat. But when it comes down to it, we need to put together regular exercise with a well-balanced diet for our own well-being.

Nutrients	Sources
Fat Soluble Vitamins	
Vitamin A	Carrots, Greens, Yellow/Orange Vegetables
Vitamin D	Dairy, Sunlight
Vitamin E	Nuts, Oils, Avocados, Tomatoes
Vitamin K	Greens, Whole Grains, Meat
Water Soluble Vitamins	
Vitamin C	Citrus, Sweet/Green Peppers, Strawberries
B Vitamins	Meat, Greens, Grains, Dairy
Folate	Greens, Liver
Minerals	
Iron	Meats, Beans, Soy
Calcium	Dairy, Soy Greens, Fish
Fiber	
Fiber	Beans, Whole Grains, Fresh Fruit

Questions about this case:

1. What are some high-fat foods?
2. What are some high-cholesterol foods?
3. Go to the following website and look up your food pyramid. What are the guidelines that you find? http://www.mypyramid.gov/guidelines/index.html
4. Why does the USDA suggest a limit on fat consumption?
5. Why does the USDA suggest a limit on saturated fat consumption?
6. What is the key disease that is associated with high-sodium diets?

Questions to go deeper:

1. What disease is high cholesterol related to? How does cholesterol participate in this disease?
2. Why is folate particularly important during **pre-natal development?**
3. What disease does vitamin C deficiency cause? How about vitamin D deficiency?
4. Iron is important for hemoglobin function. What does hemoglobin do? Why is it particularly important for women to get enough iron?
5. Where in the body is calcium a critical component?

References:

1. www.mypyramid.gov/guidelines/index.html
2. www.usda.gov

How Ed Saved the World!

A Case Study of the Small Pox Vaccine

Smallpox is clearly the worst disease to plague humans since the beginning of written history. Through a series of **epidemics** over almost 3,000 years, smallpox has shaped history by defeating seemingly invincible armies and by nearly wiping out entire people groups. By the eighteenth century, smallpox was responsible for literally millions of deaths in every known country of the world, with the exception of Australia. Some estimates put the death toll in the twentieth century at approximately 500 million. Yet today, smallpox is considered to be eradicated, which is one of the greatest success stories in the history of medicine. How this happened is as much a story of the amazing human immune system as it is of Edward Jenner, the English physician who is credited with developing the vaccine.

The story of the fall of smallpox begins with this simple scientific **observation** by Jenner in the late 1700s: while smallpox has a mortality rate of about 30%, survivors are **immune** to further infection. Interestingly, milkmaids in the English countryside who contracted cowpox (a non-fatal cousin of smallpox that primarily infects cows) were immune to smallpox, as well. You are probably thinking, "Why not just give everyone cowpox to protect against smallpox?" This is precisely what Jenner thought, but there were a few problems. First, Jenner had no scientific support for his **hypothesis.** In addition, this hypothesis was one that many people doubted. Even if he could support the hypothesis scientifically, he didn't know the most effective way to "**vaccinate**" a large number of people. The word "vaccinate" actually comes from this point in history, as *vacca* is the Latin word for cow. How might Jenner expose millions of people to diseased cows with cowpox to vaccinate them? To answer this question, Jenner did an **experiment** that would be considered unethical by today's standards. Jenner took pus from the sore of a person with cowpox and put it into a cut on the arm of a boy who had never had smallpox. After 6 weeks, Jenner then exposed the boy to smallpox. As he had hypothesized, the boy did not contract the disease. Jenner repeated this experiment on 5 other boys, including his own son, and again none of them contracted smallpox. He had supported his hypothesis and by the early 1800s tens of thousands were being vaccinated by this method.

What is in the cowpox sores that can cause such an amazing protective effect? To answer this question, we need to look at the biology of smallpox. Smallpox is caused by the **variola virus,** which is in the same virus family as the cowpox virus. The virus is usually acquired by inhalation of infectious droplets or by touching items that infected people have contacted. The virus infects the cells of the respiratory tract and once there, it essentially takes the cells over. The virus uses the person's cells to copy its own viral genetic material and to produce viral proteins. It can then assemble new viruses inside the cell. The infected cells eventually burst, releasing thousands of new viruses that travel all over the body. These new viruses create new pockets of infection, while the patient remains **asymptomatic**. Within about 14 days, the pockets of infection within the skin make deep **pustules,** which result in severe scarring for survivors. These pustules are caused by a **cell-mediated immune response**. This means that white blood cells from the immune system attempt to fight the infection by killing infected cells. The dead cells collect in big pockets, leading to deep pustules. Although this may seem weird, this response helps to limit the infection. Even with the immune response, the mortality rate for smallpox is about 30%, which is high when compared to other diseases.

As Jenner observed, survivors of smallpox are immune to future infection. How does this work? Along with the cell-mediated immune response, the human body has a **humeral immune response**. This part of the immune system is designed to identify foreign molecules, for which we use the general term "**antigens**." When the body identifies a foreign antigen, it then makes **antibodies**, which recognize the antigen and bind to it so that the body can neutralize it. When a person encounters smallpox virus for the first time, his or her immune system is considered to be naïve to those antigens. While the cell-mediated response fights the infection, the humeral response makes an antibody uniquely designed to neutralize the smallpox virus. Unfortunately, it requires time to make the antibody that can perfectly neutralize the virus. Since smallpox infections are so severe, many people die before their humeral immune response can produce enough antibodies to neutralize the virus. However, take for example a less **virulent** virus like **varicella,** which causes **chickenpox**. There is a low mortality rate associated with varicella infections. In these cases, the cell-mediated response can control the infection on its own, but the humeral immune response still uses **B-cells** to produce antibodies that can specifically eliminate the invading virus if it is ever encountered again. The body then stores away a small number of these antibody-producing B-cells as **memory cells,** so that when the antigen comes around again, it is eliminated before symptoms occur (see figure on next page). When the

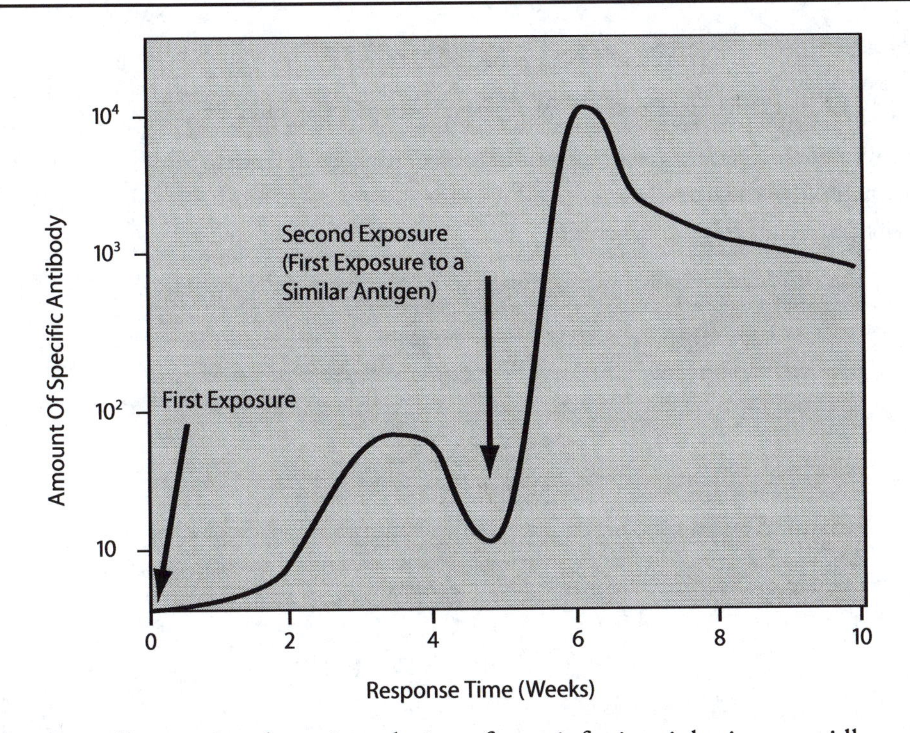

memory cell recognizes the antigen during a future infection, it begins to rapidly produce more B-cells with the ability to make the large quantities of neutralizing antibodies very quickly. These antibodies then bind to the antigen and become a tag upon which the cell-mediated response acts. The cell-mediated response can kill whatever it finds attached to the antibodies. For Jenner's work, the cowpox virus is so closely related on a molecular level to the small pox virus that a **cross immunity** is formed. The person infected by cowpox becomes immune to both cowpox and smallpox. Therefore, Jenner discovered a perfect way to vaccinate, since infection with the "harmless" cowpox can produce an antibody that recognizes smallpox. Thus, the key to creating a vaccine is to get humeral immunity to produce antibodies to a non-lethal antigen that is enough like the "real thing" that the antibody will work against the lethal antigen before it causes disease (see figure, on the next page).

Louis Pasteur later advanced Jenner's work in the 1880s by applying similar principles of vaccination to prevent both **anthrax** and **rabies**. In order to vaccinate against the rabies virus, Pasteur used a preparation of dried spinal cord from a rabbit that had died of rabies. The virus is not **viable** after drying, so while the molecular antigens still exist, the virus is not infectious. For the anthrax vaccine, Pasteur used an old culture of anthrax bacteria that had lost its virulence. This process of damaging a pathogen to make it less virulent is called **attenuation**. Pasteur's anthrax vaccine was then tested on sheep. It showed that none of the sheep died from the vaccination with the attenuated strain of anthrax, and it provided 100% protection. Pasteur's work showed that vaccination works for more than just viruses, and has the potential to work for any type of foreign invading organism. Since then, scientists have successfully vaccinated people using related **pathogens** (cowpox virus)**, dead pathogens** (Pasteur's dried rabies virus), **attenuated pathogens** (Pasteur's anthrax strain), and even molecular components of pathogens to get this desired effect.

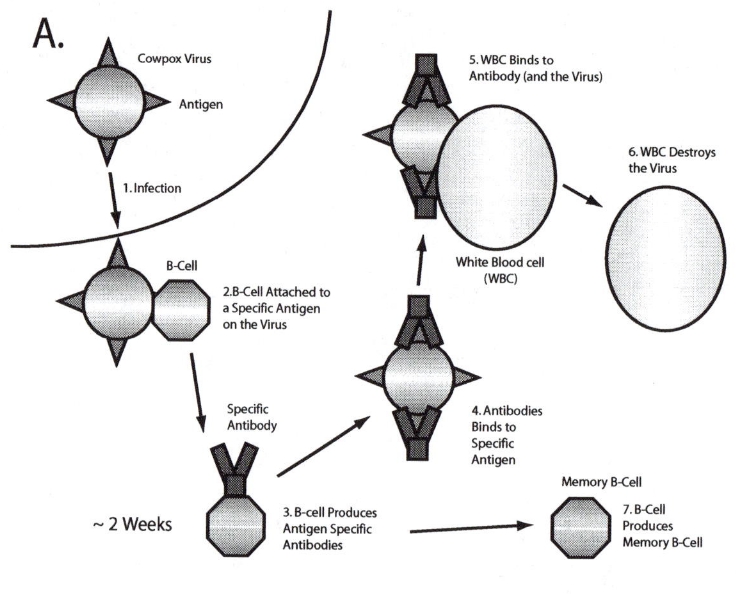

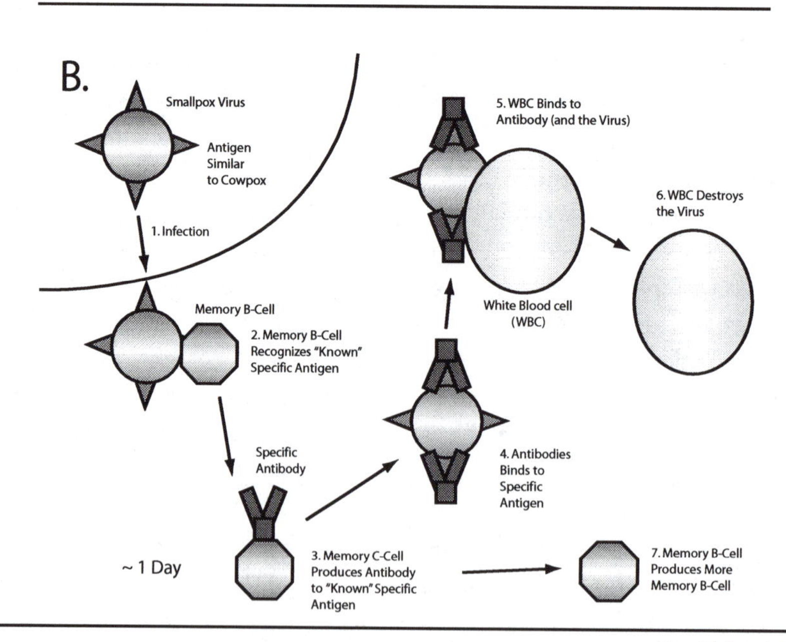

Questions about this case:

1. What was Jenner's initial hypothesis?
2. There have been cases where a vaccine has caused significant disease. How might this have happened?
3. Even when a person is vaccinated, the smallpox virus successfully infects the cells of the respiratory tract. If that is the case, when in its travels does the virus encounter the memory cell that recognizes it?
4. Influenza is also a virus. Why should we get a vaccination every year? Why is a vaccine very protective in some years and not very protective in others?
5. Why do some people get chicken pox twice?

Question to go deeper:

1. There are many scientists who say viruses are not living organisms. What would be some of the arguments that would support their claim?
2. We mentioned B-cells as part of the cell-mediated immunity. What are some of the other key components? How are these components different from B-cells?
3. Even more problematic than influenza is HIV. Attempts to make an HIV vaccine have been largely unsuccessful. Why do you think this is?
4. Some people decide not to vaccinate their children; what might be their rationale?
5. The case describes the lytic cycle of viral growth, where new viruses are made and the cell ruptures. There is also a lysogenic cycle. Describe how that works.

References:

1. Daniel A. Koplow Smallpox: the Fight to Eradicate a Global Scourge, University of California Press, Berkeley, CA, 2003, p. 1
2. Allen Chase Magic Shots: a Human and Scientific Account of the Long and Continuing Struggle to Eradicate Infectious Diseases by Vaccination, William Morrow and Company, New York, NY, 1982, p. 51

Always the Tough Guy

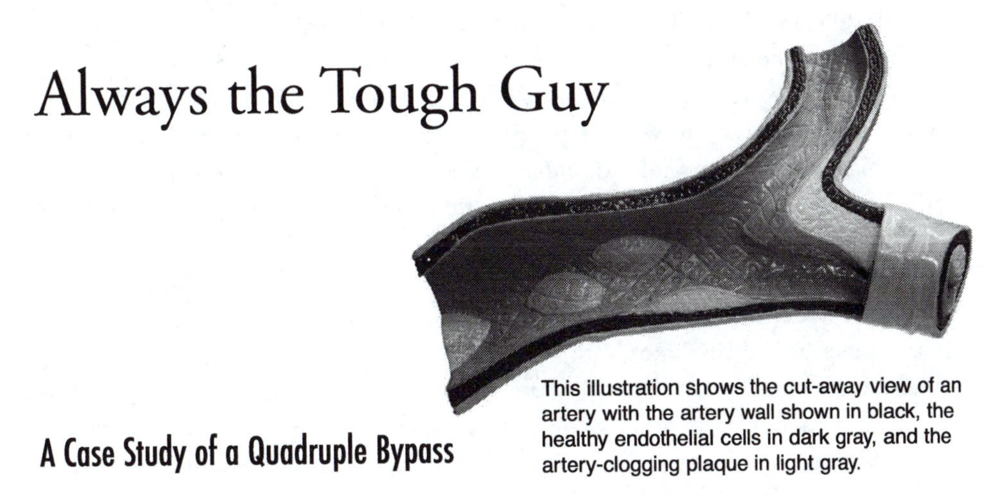

This illustration shows the cut-away view of an artery with the artery wall shown in black, the healthy endothelial cells in dark gray, and the artery-clogging plaque in light gray.

A Case Study of a Quadruple Bypass

Tim was always the tough guy, and this time was no different. A little pain in the left side of his chest and left shoulder was not a big deal. He had always fought through pain and soreness in the past. Besides, this was not major pain; it came and went. Even though the pain flared up with physical activity, Tim was gearing up for some work this particular evening. A winter storm warning had been issued, and it looked like they were about to get 12” to 24” of snow. This could amount to 2 or 3 hours of shoveling, since his snow blower quit on him during the last storm. Tim wanted to be prepared.

The storm arrived like clockwork, and it was a big one. There was no doubt that they would get 24” and Tim would probably need to shovel twice. At the 8” mark, he got his boots on and went to work on the driveway. Initially, he did not feel much pain, but 10 minutes in, the pain shot through Tim’s left shoulder and he crashed to the ground. He had never felt anything like it; he knew this was something major. Tim struggled back to the front porch and fell through the front door as he opened it. Tim’s wife, Pam, was horrified to see Tim looking as bad as he did, but stayed calm enough to call 911 and comfort Tim. The ambulance arrived in less than 3 minutes. Tim was rushed to the hospital ER, and arrived in less than 10 minutes. There was no question in anyone’s mind that Tim was having a **heart attack**. Tim couldn’t believe it; after all, guys in their 40s don’t have heart attacks, or so he thought.

In the hours and days that followed, Tim learned more than he ever wanted to know about heart attacks. He quickly realized how lucky he was to be alive. Tim soon discovered that the pain in his shoulder that he had tried to ignore was called **angina.** It was really a telltale sign that his heart was not being supplied with enough blood to meet the demand for **oxygen**. The heart is a complicated collection of muscle fibers that must contract and relax repeatedly. Like any other muscle, this cycle requires large amounts of chemical energy. The molecule that is used to supply this energy is called **ATP**, and for muscle cells to make ATP efficiently, they need to continuously break down carbohydrates in the presence of oxygen. This is why, as the heart begins to work harder, more oxygen is required to fill the energy demand. Consequently, Tim's angina would flare up only when he was physically active because at rest, his heart's demand for oxygen could be met. When Tim began to physically work, his heart needed more oxygen than could be supplied. The pain indicated that some of the cells in the heart were being damaged from lack of oxygen. However, enough of the demand was being met so he had not had a full heart attack, yet.

How is the demand for oxygen not met? Well, it can be caused by a reduced supply or an increased demand (or both). The common way people sustain an increased demand for oxygen by the heart is by physical exertion. This was what happened to Tim. But, millions of people exercise everyday without having a heart attack, so there must be other complicating factors. Indeed as in Tim's situation, many people have **progressive atherosclerosis**. There are four major arteries that supply the heart muscle with oxygen. They are called the coronary arteries. These arteries are a prime target for atherosclerosis, which is caused by the build-up of fatty deposits (called **plaques**) in the arteries. If the plaques get big enough, they can severely limit the amount of blood that can pass through the artery, therefore limiting the amount of blood (and oxygen) reaching the heart cells. If only one of the arteries is significantly clogged, then only the part of the heart that is fed by that artery is damaged. A number of factors affect how quickly these deposits grow. The common denominator for most people is high amounts of **cholesterol** in the blood. In the body, cholesterol is normally packaged with other fats and proteins into small collections called **low-density lipoproteins (LDLs).** These LDLs are specifically captured and used by a variety of cells for many different functions. However, if too much cholesterol is in the bloodstream, it will not all be absorbed by "hungry" cells; thus, there will be a high concentration in the bloodstream. Therefore, as plaques begin, the "leftover" LDLs will be absorbed into the plaque, causing it to grow. The more LDLs that there are in the blood, the more a plaque will grow. High blood cholesterol can be related to a number of factors, including genetics, diet, and physical activity. Some people just have higher cholesterol than others, but everyone can reduce their blood cholesterol by reducing dietary intake of cholesterol and **saturated fats**. Maintaining a healthy weight and exercising regularly can also help reduce cholesterol levels.

When Tim arrived at the hospital, the physicians did a simple blood test, which showed high amounts of troponin I, creatine phosphokinase, and lactate dehydrogenase in the blood. These enzymes are normally found in low levels in blood and high levels in heart cells. When heart cells are damaged by the lack of oxygen, these enzymes leak out into the bloodstream. Once Tim was stabilized in the ER, the physicians performed an **angiogram**. This is when a dye is injected into the blood, so that the blood flow through the coronary arteries can be checked. Depending on what the angiogram indicated, there were a couple of options to help Tim. First, a simple **angioplasty** could be performed. In this procedure, a device with a small, deflated balloon would be inserted into one of Tim's arteries and fed into his coronary arteries. When the device is positioned at the point where the plaque is thickest, the balloon would be blown up, thus pushing aside the plaque and making more room for blood to flow. Angioplasty is a temporary solution, and people who have had serious heart attacks often require more serious intervention. Second, physicians can surgically insert a tube through the plaques so that blood can pass through them. These tubes are called **stints**. Finally, surgeons can do a traditional **bypass** where blood vessels from the patient's leg are removed and connected to the coronary artery on each side of the plaque, therefore allowing the blood to travel down the coronary artery (but "bypassing" the plaque).

Tim's angiogram clearly showed that one coronary artery was almost completely closed and the other three were in very bad shape. It was quickly decided that performing a bypass on all four coronary arteries would be the best solution. The **quadruple bypass surgery** went well, and it was determined that Tim's heart sustained only minor damage from the heart attack. There was every indication that he would fully recover.

During his time in the hospital, Tim learned about what had happened to him and how to prevent it from happening again. He learned basic things, like why medical professionals called heart attacks "**myocardial infarctions**" (MI). This term is simply a technical way to say that heart tissue was damaged from lack of oxygen. Tim also learned that it was not uncommon for a 48-year-old to have a heart attack. Tim discovered that his blood cholesterol level was 240 mg/dl. He was also told this was high, but had no idea what that really meant. This number represents the concentration of total blood cholesterol. The number comes from adding together the amounts of LDLs and **HDLs** (high density lipoproteins) in the blood. Earlier we have mentioned LDLs; however, we failed to mention that they are often called "bad cholesterol." The reason for this is because LDLs have much more cholesterol and contribute more to atherosclerosis than HDLs. Since, HDLs have less cholesterol than LDLs and high levels are considered good, these are called "good cholesterol." Normally, total blood cholesterol should be kept below 200 (with LDL below 130 and HDL above 40). It turns out that Tim's total cholesterol level of 240 mg/dl was extremely high, especially considering that his LDLs were at 190 mg/dl and his HDLs were at 50 mg/dl.

Tim was destined for a change in lifestyle. His physician put him on a strict diet that was low in cholesterol and saturated fat. The doctor also recommended a regular exercise regiment. The diet really hurt. Tim loved Pam's southern-style cooking, but they both realized these favorites would need to be reserved for special occasions. The exercise scared Tim; after all, wasn't he essentially exercising when he had his MI? His physician assured him that regular moderate exercise was key to keeping his weight down and his heart healthy. Finally, Tim went home with a prescription for a drug to help reduce his LDLs.

Tim still gets regular check-ups that include blood tests. His cholesterol is down to 175, he has lost 20 pounds, and he exercises regularly. Tim will readily admit that the worst part about his heart attack is that he can no longer eat his wife's biscuits and gravy on a regular basis.

Questions about this case:

1. What would be some good, specific dietary recommendations for Tim?
2. How would a diet that contained no cholesterol affect a person's blood cholesterol levels?
3. What is familial hypercholesterolemia? What can we learn about the relationship of LDLs to MIs from it?

Questions to go deeper:

1. Atherosclerosis can also cause a stroke. How are MIs and strokes similar?
2. Why are people who are at risk for a heart attack often told to take an aspirin a day?
3. Why are some heart attack patients prescribed digitalis?

From Depression to BoTox®

A Case Study of the Action of Neurotransmitters

When we met Jennifer, it did not take long to figure out something was wrong. Jennifer was a difficult person to talk to; she did not seem to want people to know much about her. Most questions she just answered with a yes or no, and she never seemed happy. As we got to know Jennifer better, she would share information about herself. It became obvious that she was very bright and incredibly talented. There were times when she was cheerful and lots of fun, but there were lots of days when she seemed depressed. "Depressed" is a layman's term for feeling down or blue, but the overuse and misuse of the term has led to significant misconceptions. Jennifer was indeed depressed, **clinically depressed**. She had lived in a community that made her believe that she should just be happy. She would get even more depressed at the notion that she could not control what it seemed everyone else could. Famous movie stars would be on the television saying that depression is not real and that depressed people should just take some vitamins and be happy. She was embarrassed and felt weak because she could not handle simple events in life without being engulfed by depression. Jennifer did not seek help for many years. After all, she was not suicidal, she just routinely felt sad. She could hardly interact with people she knew, much less talk with people she did not. Being convinced that nothing was wrong kept her from seeking professional help. The turning point for Jennifer was becoming good friends with a person who had a recent diagnosis of clinical depression.

It took Jennifer many months to realize that depression is a disease. Jennifer did not understand that the people around her that seemed to handle life better than her were not struggling with depression. When Jennifer finally understood that depression involves a chemical imbalance in the brain that vitamins couldn't fix, she made an appointment to see a physician. Her physician made it very clear that depression takes many forms, from mild anxiety around crowds to suicidal feelings, yet the whole spectrum of symptoms have a similar source. To better understand the chemical basis of depression, Jennifer's physician sat her down and explained how the brain sends and receives messages.

When the brain sends a message, it travels through brain cells called **neurons** as an electrical current. Neurons have long appendages called **axons** that extend to the location where the message needs to be delivered. The axon is much like an insulated wire in your home that carries electricity from one place to another. There is a small gap between the end of the axon and the **dendrite** of the neuron that receives the message, called a **synapse**. This gap prevents the electrical message from being transmitted from the axon to the dendrite. Instead, the neuron produces chemicals called **neurotransmitters**, so when the electrical message reaches the end of the axon, the neurotransmitter is released. The dendrite has receptors for the neurotransmitter and when the neurotransmitter finds the correct receptors on the dendrite, the message is converted back into an electrical current. After this occurs, the neurotransmitter is released and travels back to the axon to be taken back up into the axon and reused. Patients with depression have been found to have decreased amounts of the neurotransmitter **serotonin** in parts of their brain responsible for mood. Correspondingly, the most common drug treatments for depression have their effect on serotonin. **Prozac**, for example, is the best known member of a group of drugs called selective serotonin reuptake inhibitors, or **SSRIs**. As their name implies, these drugs prevent serotonin from being taken back up by an axon after it is released from the receptor. As a result, there is more serotonin out in the synapse and a greater chance that serotonin will bind to a receptor on a dendrite. The result is an increase in messages picked up by the dendrites of the neurons involved in establishing mood.

Jennifer was prescribed an SSRI for her depression and after a bit of trial and error, a dose was determined that had minimal **side effects**. She has since found her depression much more manageable.

Since neurotransmitters transmit messages from both neurons to neurons and from neurons to muscles, there is a huge variety of diseases in which they are involved. For the same reason, there are a variety of disease treatments that involve neurotransmitters, as well. To illustrate this, we will look at a potentially life-threatening infection that seems completely unrelated to depression, yet involves neurotransmitters.

If you have any experience with home canning, you have almost certainly heard of **botulism**. Botulism is a potentially fatal disease caused by the bacterium ***Clostridium botulinum***, which is found in soil and on many types of vegetables. The bacteria itself is not the serious problem, but the toxin it produces is. **Botulism Toxin type A** is one of the most toxic substances known to man, and there are three ways people come into contact with it. In adults, botulism is usually a food poisoning, which has been most often acquired from improperly canned food. Improperly canned food provides an excellent environment in which the bacteria can grow. *C. botulinum* germinates in an **acidic pH** in an **anaerobic** environment. This is precisely what you find in home-canned acidic foods like tomatoes that have not been sterilized. In this situation, an adult ingests the toxin that was produced as the bacteria grew in the food. In an infant, botulism usually takes the form of an intestinal infection by the bacteria. As the bacteria grow, they produce the toxin, which can have very serious consequences. It is also possible to have an open wound infected with the bacteria and, as in the intestinal infection, the toxin is produced as the infection spreads. There are about 100 cases of botulism poisoning per year in the United States. Of the 100 cases, about 70% are infants, about 30% are food poisoning, and cases of wound infection are rare (1). The symptoms for botulism toxin poisoning all stem from the ability of the toxin to cause **flaccid paralysis**. This simply means that it prevents muscles from contracting.

This toxin can have catastrophic effects. If muscles cannot contract, then the muscles of the **diaphragm** cannot function and breathing becomes difficult or impossible. How does this toxin cause such a monumental problem? As in our example of depression, the answer lies in how neurotransmitters work. When the brain sends a message to a muscle to tell it to contract, the electrical message travels down an axon to the point where it meets the muscle. There is a small gap between the end of the neuron and the muscle that is much like the synapse we mentioned earlier, but is called a **neuromuscular junction**. Like the synapse, the electrical message cannot cross the neuromuscular junction. When the electrical message reaches the end of the neuron, a neurotransmitter called **acetylcholine** is released. The acetylcholine travels across the gap to bind receptors on the muscle cell. This tells the muscle to contract. The neurotransmitters are made by the neuron and sent to the **plasma membrane** at the end of the axon, in small packages called **vesicles.** The vesicles are released when the electrical message arrives. Botulism Toxin type A prevents those vesicles containing acetylcholine from spilling their contents into the neuromuscular junction. Without acetylcholine binding to its receptor, the muscle will not contract. The action of the toxin is difficult to reverse, so in severe poisonings, patients who survive require respirators, and recovery is very slow.

If Botulism Toxin type A is one of the most toxic substances known to man then why would anyone consider purposefully coming into contact with it? In the 1980s and 90s the toxin was found to be helpful in treating a number of disorders that involved localized muscle tension. Soon thereafter, research indicated that it may have cosmetic use for easing facial wrinkles due to aging, since these wrinkles are a result of muscle tension. If the facial muscles could relax, the wrinkles would be reduced, and the patient would look younger. As a result, by 2003, the injection of Botulism Toxin type A in people's faces under the trade name **BoTox®** had become the most popular non-surgical cosmetic treatment in the United States (2). Treatments are localized injections of diluted Botulism Toxin type A, which can remove wrinkles for 3 to 4 months. At an average cost of more than $400 per treatment and with more than 2 million treatments performed annually, this has become a billion-dollar industry, without even considering the BoTox® cosmetic line (2).

From these examples, it should become evident that diseases and treatments involving neurotransmitters can be quite diverse.

Questions about this case:

1. If axons are much like insulated wire, then what insulates them?
2. In general, what are side effects and what are the specific side effects for SSRIs.
3. Sarin, a potent neurotoxin, was released into the Tokyo subway system by terrorists in 1995. How does Sarin affect neurotransmitters?
4. How does nicotine affect neurotransmitters?

Questions to go deeper:

1. Myasthenia gravis is an autoimmune disease involving neurotransmitters. Briefly describe the problem in this disease.

References:

1. www.cdc.gov/ncidod/dbmd/diseaseinfo/botulism_g.htm
2. www.surgery.org

A Fight for a Family

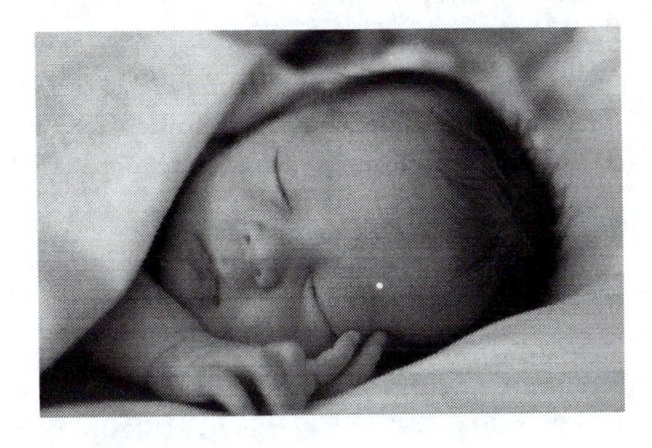

A Case Study of Hormones, Pregnancy, and Miscarriages

As a biologist, I too am impacted by biological principles in my daily life. I wanted to share one such occasion that hit particularly close to home.

When I married at age 27, my wife, Tammy, and I decided to wait a couple of years to try to have kids. When we began to try to get pregnant, it did not happen as easily as we thought it would. Month after month went by without Tammy getting pregnant. Tammy and I are both biology teachers, yet neither of us was an expert in pregnancy. Tammy is quite bright and resourceful, so she began to research why women fail to get pregnant and what they do to give themseves a better chance of getting pregnant. We both knew that the female **menstrual cycle** lasts approximately 28 days, with **ovulation** occurring at about day 14. What we did not know was that there was a small spike in body temperature the day of ovulation that can be seen if you take your temperature every morning when you wake up. The goal was to know when ovulation occurred, so we would know when an **egg** was available for **fertilization**. It got to the point that Tammy knew the day she ovulated almost without taking her temperature every morning, and after 12 months of trying, we got pregnant! All the excitement came crashing down 7 weeks later when Tammy began bleeding. We feared that we were having a **miscarriage**, and an appointment with our physician confirmed it. After the medical exam, the physician determined that it was not necessary to perform a **D and C**. The abbreviation D and C stands for dilatation and curettage. In this procedure, the **cervix** is dilated and a small instrument is inserted to help remove any of the material left in the **uterus**. Since Tammy had passed the whole **embryo** during the bleeding, a D and C was not required.

We were crushed, but began trying to get pregnant again. Tammy also continued her research online. Tammy knew that her cycle was consistently shorter than 28 days. In fact, her cycle was usually around 21 days. What she learned was that this could be the source of the problem. The menstrual cycle is broken into two phases, the **follicular phase** and **luteal phase**. In the follicular phase, the **pituitary gland** produces **lutinizing hormone** (LH) and **follicle stimulating hormone** (FSH). These hormones cause an egg-containing "**follicle**" within the **ovary** to grow. During this time, the lining of the uterus thickens as it collects blood in preparation for embryo implantation. This phase ends with a spike of LH and FSH, resulting in ovulation. During the luteal phase, the follicle that ruptures to release the egg turns into a **corpus luteum,** which produces large amounts of the hormone **progesterone**. Progesterone is responsible for the maintenance of the lining of the uterus so the growing embryo can implant properly. After the egg is released from the ovary, it travels down the **fallopian tubes**, becomes fertilized by joining with the incoming **sperm**, and travels into the uterus where it finds a suitable environment to imbed into the lining. In women with a short menstrual cycle, it is common to have a normal 14-day follicular phase and a short luteal phase. This is sometimes referred to as a luteal phase defect.

The problem with the short luteal phase is that progesterone may not be made in high enough amounts to properly maintain the uterine lining; therefore, implantation of the embryo may not be successful. Through her research, Tammy found she had a luteal phase of about 7 days, which was considered a significant luteal phase defect. Tammy made an appointment to consult with her physician, but found out two days before the appointment that she was pregnant again. Thinking Tammy no longer needed the appointment, her physician cancelled it. Tammy was angry and desperate. If her short luteal phase was the problem, then she desperately needed that appointment. Tammy pressed the issue until she got an appointment with an **ob/gyn** specialist. At the appointment, Tammy explained what she had learned without trying to sound like she was telling the specialist what to do. Oddly enough, her specialist was married to a woman with the same luteal phase defect and was a father of three. During the pregnancies of each of his children, his wife was given supplemental progesterone. Tammy was prescribed progesterone, which I administered as a shot once a day for the first 3 months of the pregnancy.

Human **gestation** is about 40 weeks, which translates into about 9 months. This is divided into **3 trimesters** of about 13 weeks each. The first trimester is where the most miscarriages occur, yet we were nervous during the whole pregnancy due to our previous experience. As the due date arrived, it was clear to our physician that the baby was in the **breech** position. We made a special appointment to try an **aversion**. This is where muscle relaxants are given to try to relax the uterus and then the doctor physically tries to turn the baby around. In our case, this was unsuccessful. At 39 weeks, it was determined that our baby was healthy, but that Tammy was no longer making enough **amniotic fluid**. We were told that a **Cesarean section** (C-section) was the best option, and we planned it for the following Monday morning. To prepare for the C-section, Tammy was given a **spinal block**, which means that a catheter was inserted between two spinal vertebra in her back into the **epidural space** (not into the **spinal cord** itself) through which a strong **anesthetic** was administered. This action deadened the pain from her chest to her toes. This block is different than the **epidural** given during a **vaginal birth**. An epidural is given in the same manner and location, but it does not completely block the activity of the spinal cord. With the epidural, pain is significantly reduced, but it does not prevent the control of the uterine muscles so the "mother-to-be" can still have a vaginal birth. When it was established that Tammy could no longer feel pain, the surgeon then made a transverse incision through the skin and uterine muscle layers to remove the baby. The incision was small and low on her stomach, so metal retractors were inserted to help open the incision and a second physician pushed down hard on the baby so that he was forced through the opening. Our son, Justus, was born June 15, 1998.

We successfully got pregnant again a little over a year later. Again, with the help of supplemental progesterone (this time oral, no injections!), Tammy carried our daughter, Jenna, to term. As the due date arrived, we were counseled on the possibility of having a **VBAC** or a **v**aginal **b**irth **a**fter **C**esarean. Many people assume that after you have one C-section you can not have a vaginal birth. We were told that this was not the case and that we should consider the VBAC. Unfortunately, our daughter was also breech, and we went ahead with a second C-section.

A happy ending, right? Of course, but this is not Hollywood. We got pregnant 3 years later. This time it was not planned. We had since moved to a new city and had not established a relationship with a physician or a specialist. With the slow process in getting appointments and referrals with **HMOs**, we miscarried before we could get an appointment to see a specialist. But this experience taught us some valuable lessons. First, when it comes to health care, we need to advocate for ourselves. There was no one in the health care system to make a clear case for that first appointment with the ob/gyn specialist we desperately needed. And second, Tammy's personal research was a valuable tool. The Internet is full of reputable sources of information in every field of biology. Sure, there are always some quacks out there, but if you use care, you can find reliable information on almost any topic. It's my hope that you can become comfortable enough researching biological topics that when you encounter situations like mine, you can find good answers to your questions.

Questions about this case:

1. What does ob/gyn stand for? What do these specialists do?
2. How do oral contraceptives work?
3. What is the difference between an embryo and a fetus?
4. What is an episiotomy? When is it used?

Questions to go deeper:

1. What are some other female-specific physiological reasons that might make getting pregnant difficult? What are some male-specific physiological reasons?
2. In what other occasions are D and Cs used?
3. How does the muscle of the uterus compare to skeletal muscle?
4. Describe the system through which the developing baby receives nutrients and eliminates wastes.

Laura Wishes She Only Had Mononucleosis

A Case Study of Leukemia and a Bone Marrow Transplant

Laura was tired, not just tired, but really tired. Initially, it was no big deal. Everyone gets worn-out, and her friends knew she would bounce right back. She was the one who was always on the go, so after a couple of weeks her friends and family were starting to get worried. The general consensus was that she had a virus, but some folks thought maybe she had mononucleosis. Either way, Laura eventually became convinced that she needed to see her physician.

Dr. Baker came in with her chart. "So what brings you in today, Laura?"

"It's just not like me," Laura began. "I am the one with all the energy and now I am just so tired, I am afraid I might have mono."

"How long have you felt this way?" Dr. Baker continued.

"Almost two weeks now," said Laura.

As the conversation continued, Dr. Baker felt Laura's neck, looked into her ears and mouth, and listened to her breathing.

"Have you had any other symptoms like fever, joint pain, or weakness?" asked Dr. Baker.

"Now that you mention it," Laura said, "I have had some odd pain in my arm, almost like the bone hurts, right here," Laura said as she showed her the spot. As Dr. Baker examined it, she said, "I am going to send you down to the lab to have some blood drawn for a couple of routine tests. For now you go home and rest."

A couple of days later, Dr. Baker's office called and asked Laura to come back in for a follow-up appointment to go over her blood work. Dr. Baker's receptionist seemed very calm, but Laura felt uneasy. They made an appointment for the next day. At the appointment, Dr. Baker explained to Laura and her parents that Laura's **white blood cell** (WBC) counts were unusually high and that she had made an appointment for Laura with Dr. Williams, whose specialty was **hematology** and **oncology**.

At this point, Laura was scared. She knew enough to realize that oncology referred to cancer. When they met with Dr. Williams, she very calmly explained that high WBC counts could be a symptom of many different problems, including **mononucleosis**. Dr. Williams went on to explain that the fatigue, high WBC counts, and bone pain all can be symptoms of **leukemia**, which is **cancer** of the blood cells and **bone marrow**, but more tests were needed to determine the correct diagnosis. Laura had gone from scared to terrified, and the 3-day wait for the test results was agonizing. When she and her parents sat in Dr. Williams' office, they knew the news was not good. "The blood tests show that you have **chronic myelogenous leukemia,** or CML for short," Dr. Williams said. "We still need some additional tests, but we are fairly certain this is the correct diagnosis." Laura and her parents were devastated. After sitting quietly for a minute, Laura gathered herself and began to ask some questions. "What is CML," she began, "and what does this mean? I mean, how bad is it, and is it treatable?" "Those are great questions and I want to spend the time to explain this completely," Dr. Williams said.

CML is a **malignant** cancer of cells that live in the bone marrow. Bone marrow is the semi-liquid core of larger bones that contains a type of stem cell. As the stem cells divide and develop, mature **red blood cells** (RBCs) and a variety of types of WBCs are produced. When one of these stem cells becomes malignant, it divides more rapidly than it should and the bone marrow gets crowded with immature WBCs. Eventually, these cells get released into the blood, even though they are immature and not functional. The result is a high WBC count. This particular cancer occurs in two phases. First, there is a **chronic phase**, where the cancer cells grow slowly and there may not be any symptoms. The chronic stage is treated with **chemotherapy**, which can result in temporary **remission**. The second phase is the **acute phase**, where the cancer cells grow very quickly and produce many immature blood cells called "**blasts.**" It can have quite dramatic symptoms. The acute phase is very serious and can progress to what is called blast crisis. Traditional chemotherapy has little effect on acute phase CML, so the best treatment is a **bone marrow transplant**.

"Chemotherapy" is a general term for any chemical treatment, but is specifically used regarding drug treatments for cancer. These drugs generally kill or injure cancer cells by targeting cells that are dividing rapidly. Unfortunately, there are many normal cells that divide rapidly, such as hair follicle cells and cells that line the intestinal tract, that would also be killed or injured by these drugs. This results in the common side effects of hair loss and severe nausea and vomiting.

Bone marrow transplants are a much more involved process than simple chemotherapy. This procedure removes the existing cancerous cells and replaces them with healthy donor marrow. But as in any transplant, finding a match can be difficult because the donor marrow will need to be compatible with the existing immune system. This new marrow must come from a person that is so similar to the recipient that the existing immune system does not react as if the donor marrow is foreign and try to kill it. If the donor and recipient were not properly matched, the recipient's immune system might make antibodies to direct the killing of donor marrow, and this would result in transplant **rejection**; likewise, the donor marrow also could recognize some molecules in the patient as foreign and begin to fight them. This would be called **graft versus host disease**. The matching process includes much more than just simply matching blood type, but a potential donor needs to only submit swabbing from the inside of the cheek for a lab to determine if they are a match. Siblings are the most likely marrow donors, otherwise a donor might be found through the **National Marrow Donor Program**. Once, if, a suitable donor is found, a very large needle is used to bore into a marrow-containing bone, usually at the crest of the hip, and the fatty semi-liquid bone marrow is sucked out and stored until it is used. The transplant recipient is treated with chemotherapy to kill the cancer cells and suppress the immune system to prevent rejection. The marrow is then given by IV, and as it travels through the bloodstream, it will naturally take up residence in the bone marrow. It takes about 2 weeks for the donor marrow to produce sufficient blood cells. Until that time, the recipient must be kept in a sterile environment to prevent an infection that the immune system would have to fight off. The recipient must also get blood transfusions to supply the blood cells that the marrow is not yet making.

CML is 100% fatal if not treated, and the average life expectancy without a marrow transplant is about 4 years. However, the **prognosis** with a successful marrow transplant is that more than 50% of patients survive 5 years or longer. Laura's CML is starting into the acute phase, so her process will begin with the search for a bone marrow donor.

Questions about this case

1. Which bones make blood cells?
2. In relative terms, how often do mutations result in a selective advantage?
3. How are benign and malignant tumors different?
4. What are the dangers of a bone marrow transplant?
5. How do the living cells in the bone marrow get their blood supply?

Questions to go deeper

1. How might a cell go from being normal to malignant?
2. Are benign tumors in the process of becoming malignant? Explain.
3. Can benign tumors be life threatening? Explain.

Reference

http://www.nlm.nih.gov/medlineplus/ency/article/000570.htm

A Poor Way to End a Great Vacation

A Case Study of Myasthenia Gravis, an Autoimmune Disease

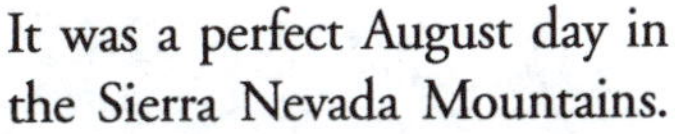

It was a perfect August day in the Sierra Nevada Mountains. Jon had backpacked annually with his friends in this area for almost 25 years, and it was always the highlight of his summer. This trip was no different. He was leading one of 10 groups of college students into the high country, and he was looking forward to a little reading and fishing.

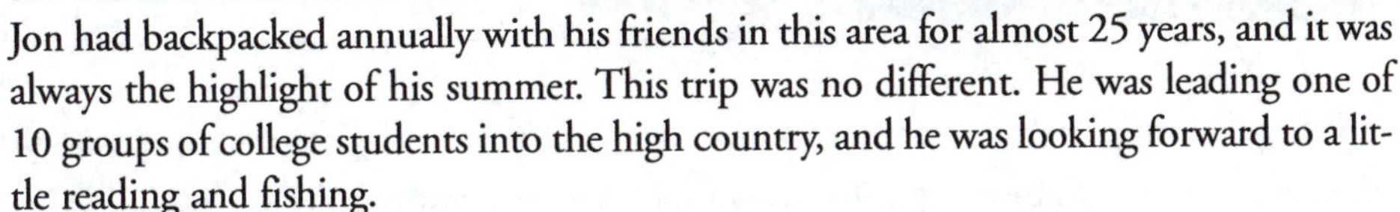

The problems began about 3 days in. It started simply with blurred vision and headache. Of course, Jon didn't tell anyone; he was the leader of the trip, and he didn't want to be a burden on others. The problem was that the symptoms quickly progressed so that he could no longer hide them. The blurred vision got worse and his speech began to slur. The decision was made that two hikers would help lead Jon out of the mountains so he could be seen by a physician.

Dr. Reed read the chart. A 45-year-old Caucasian male with a sudden onset of blurred vision, headache, and slurred speech. All vital signs taken by the nurse seemed normal. Jon was not taking any prescription medicines, and his personal and family medical history gave no clues. Dr. Reed was aware that there were many things that could cause a rapid onset of these symptoms. Something as simple as **hypoglycemia** could be the problem, so he ordered routine blood tests. These tests came back negative, indicating the diagnosis would be more complicated. Jon went through CT scans, MRIs, more blood tests, and visits to specialists. Fortunately, **multiple sclerosis** (MS) was ruled out, but the evidence seemed to point to a related disease called **myasthenia gravis** (MG).

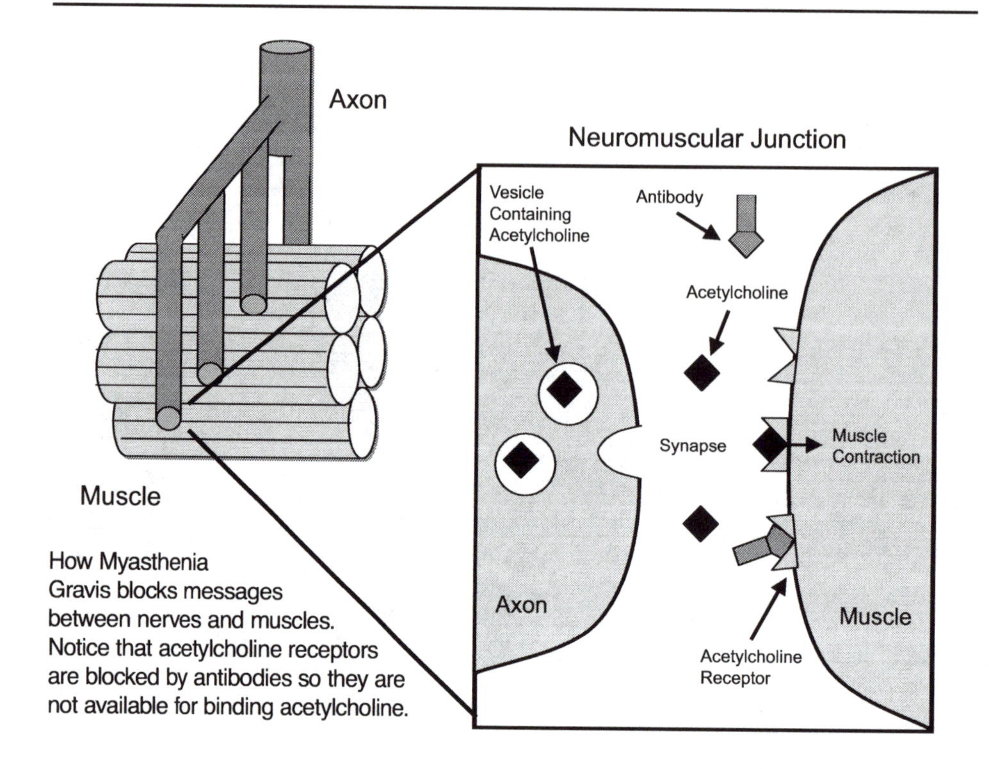

How Myasthenia Gravis blocks messages between nerves and muscles. Notice that acetylcholine receptors are blocked by antibodies so they are not available for binding acetylcholine.

While finally having an answer was a relief, Jon had never heard of MG, so he had no idea what this all meant. Jon began with a series of questions. What is MG? How will it progress? Is it treatable? How normal will my life be? All very reasonable questions, but the answers would require some understanding of how the brain sends messages to muscles and how the immune system works.

The slurred speech and blurred vision that he was experiencing were the result of the loss of fine muscle control. Normally, the brain sends out electrical messages to these intended targets in order to control them. **Neurons** conduct these messages. Neurons are highly specialized cells that are made up of three significant parts: **axons, dendrites**, and the **cell body**. Dendrites act as message receivers, axons as the message senders, and the cell body keeps the neuron alive.

When a message originates in the brain, it must be sent through a series of neurons, since a single neuron is not long enough to reach from the brain to the intended muscle targets. From the point of origin, the electrical impulse is carried down an axon until it reaches the next neuron in the chain. Axons are able to conduct the electrical impulses effectively because they are insulated with a protective **myelin sheath**. This basically allows the axons to act as insulated wires to carry the electrical messages.

Once the electrical message reaches the end of the axon bundle, it is converted into a chemical message called a **neurotransmitter,** so it can be transmitted to the next set of neurons in the chain. Neighboring neurons do not touch, but they are separated by a small gap called a **synapse**. The chemical message is then received by the dendrites of the neighboring neurons.

Once the chemical message is received, it is then reconverted into an electrical impulse, and the process begins again. When the messages arrive at the muscles, they must be transferred to the muscles in the form of a neurotransmitter, since the neuron and muscles cells also do not physically touch. They are separated by a small gap called a **neuromuscular junction**. When the electrical message reaches the neuromuscular junction, the neuron releases more neurotransmitters, like **acetylcholine**, that serve as a way to bring the message from the neuron to the muscle cell. **Receptors** on the muscle cells can receive this chemical message and translate it into muscle contraction.

MG is an **autoimmune disease**. This simply means that Jon's immune system is actually attacking some normal part of his body. For reasons that are unknown, Jon's immune system has determined that the receptors on the muscles cells that receive the neurotransmitter are "foreign." His immune system has directed white blood cells, called **B-cells,** to make antibodies that will bind to them so they can be destroyed. Muscle weakness occurs when a message travels down a neuron to a neuromuscular junction, causing the neurotransmitter, acetylcholine, to be released. When acetylcholine travels across the gap to find a receptor on the muscle cells, some portion of the receptors are unavailable because antibodies are binding to them. With fewer acetylcholine molecules binding to receptors, there is a weaker muscle contraction. Since the amount of antibody produced increases over time, the first symptoms usually show up where very fine control of muscles are required like in the eyes and tongue. This resulted in Jon's blurred vision and slurred speech.

Dr. Reed informed Jon that MG cannot be cured, but can be controlled with **steroid** treatments. Steroids are **hormones,** which are just molecular signals that tell different cell types to do different things. Some cells respond to steroids by breaking down sugars to make more ATP; others may build up muscle fibers. Oddly enough, cells on the immune system die when high amounts of steroids are present. It is the inappropriate activity of these cells that causes the problem to begin with, so this treatment aims to kill enough of these cells so the symptoms disappear without weakening the immune system too much. Steroid treatments are common in many diseases that are a result of an overactive immune system like severe allergies and asthma.

Like MG, multiple sclerosis (MS) is also an autoimmune disease that is related to muscle contraction and is often treated with steroids. However, the symptoms of MS are often much more severe than MG. MS actually has a variety of symptoms that include muscle weakness, lack of coordination, visual and sensation problems, difficulties with speech, severe fatigue, and pain. In severe cases, MS may be debilitating.

MS is caused when the immune system attacks the myelin sheath surrounding axon bundles. This attack leads to the death of the **glial** cells that form the sheath. When this occurs, it has the same effect as when you remove the insulation from a wire: electrical signals are no longer effectively conducted. Therefore, as the glial cells are destroyed, the axons that they surround lose their "insulation," so they can no longer transmit electrical messages from the brain to the desired target. As a result, the brain loses communication with that area of the body, and the muscles in this area no longer respond. This leads to weakness and lack of coordination in that area of the body.

Global attacks on the nervous system usually lead to debilitation disease because the muscles throughout the body are no longer "in contact" with the brain. The immune system can also attack the glial cells surrounding the neurons that carry sense information, leading to a lack of sensation for the sufferer. Like MG, MS is also treated with steroid treatments, but even with the treatments, rapid progression, leading to debilitation, is rather common.

Jon is fortunate. His MG is not progressing rapidly and the steroid treatments are working well. Most days he does not notice he has MG. Sometimes, stress or lots of strenuous activity causes the symptoms to return, and his physician simply adjusts the steroid prescription. Jon still backpacks every August; he just doesn't carry as heavy a pack as he used to.

Questions about this case:

1. What other cells in the immune response, besides B-cells, are important in the immune response?
2. What are some of the other symptoms of MG and MS?
3. If MG is allowed to progress unchecked, it can result in death. What could be the cause of death?

Questions to go deeper:

1. How does a neuromuscular junction differ from a synapse?
2. Cells of the immune system die when treated with steroids in a process called **apoptosis**. What is apoptosis?
3. Since steroids signal many types of cells while treating a disease, steroids invariably have side effects. What are some of the side effects of treating MG or MS with steroids?

Pulling an All-Nighter

A Case Study of Caffeine and Signal Transduction

It's final exams week, and for most students, the studying has been quite intense. There is too little time to finish all the projects and study for the exams. Nate is a business major and his finals schedule is particularly intense. There just seems to be no way to prepare for his macroeconomics and marketing finals, which are back to back on Wednesday morning, and finish a major paper for his literature class. Nate hates to pull "all-nighters," but it looks like that might be necessary for Tuesday night. He studies most of Tuesday, and by late evening, he begins to work on the paper. As it gets late, Nate realizes that he is going to need some help staying awake. He has planned ahead and has a refrigerator stocked for that purpose. As the evening turns to early morning, Nate has made good progress and has stayed alert by consuming four energy drinks, all of which are very high in **caffeine**. Having made such good progress, Nate decides to get about 3 hours of sleep, so he slips into bed and sets the alarm. As Nate lies there, he finds relaxing difficult. His heart is beating so hard and fast that he thinks something is wrong. He can hear his heartbeat in his ears and his chest is pounding. Fortunately, the next thing Nate remembers is his alarm going off. Nate hurries off to class with his heart beating much more normally. His studying pays off in two excellent performances on his finals.

That evening, Nate begins to wonder if the abnormality of his heartbeat early that morning was just a dream. To answer that question, Nate consults a biology text regarding the physiological action of caffeine. As Nate reads, he realizes that caffeine alters the way some cells process the messages they receive. Cells of your different organs send messages by making and releasing molecules that serve as signals. Cells receive messages by making **receptors** for the specific signal molecules to which they respond. This signaling can take three forms: **endocrine**, **paracrine**, and **autocrine**. Endocrine signaling happens when a signal molecule is made by a tissue and sent in the bloodstream throughout the body to affect cells that have the appropriate receptor. An example of endocrine signaling is **estrogen**, which is made in the **ovaries** and affects many cell types. Paracrine signaling is the same as endocrine, except the signal molecules are used to affect only local cells. Autocrine signaling involves a cell producing a signal that affects itself. An example of paracrine signaling is the production of **neurotransmitters** by one neuron to affect the adjacent neuron. Some **cytokines** are produced by the same cell they act on, which would be an example of autocrine signaling. Most of the receptors are confined to the **plasma membrane**. When receptors on the plasma membrane bind to their specific signal molecule, they pass the message over by activating some other molecule, which in turn, passes the message over by activating another molecule. This results in a "cascade" of newly activated molecules, until the message has its final effect. This is often called a **signal transduction cascade**. Generally, the signals result in either altering the function of a protein (by modifying it) or altering the amount of a protein by changing the **transcription** of the gene that codes for it. Either way, the cell becomes functionally different by the binding of a molecule to a receptor in the plasma membrane. Any molecule that is in the organized cascade is called a **second messenger**. A specific signal molecule may find the same receptors on a variety of different cells and in binding to those receptors, may activate a similar signal transduction cascade, but with a different final result. For example, **adrenaline** (also known as epinephrine) is an endocrine hormone

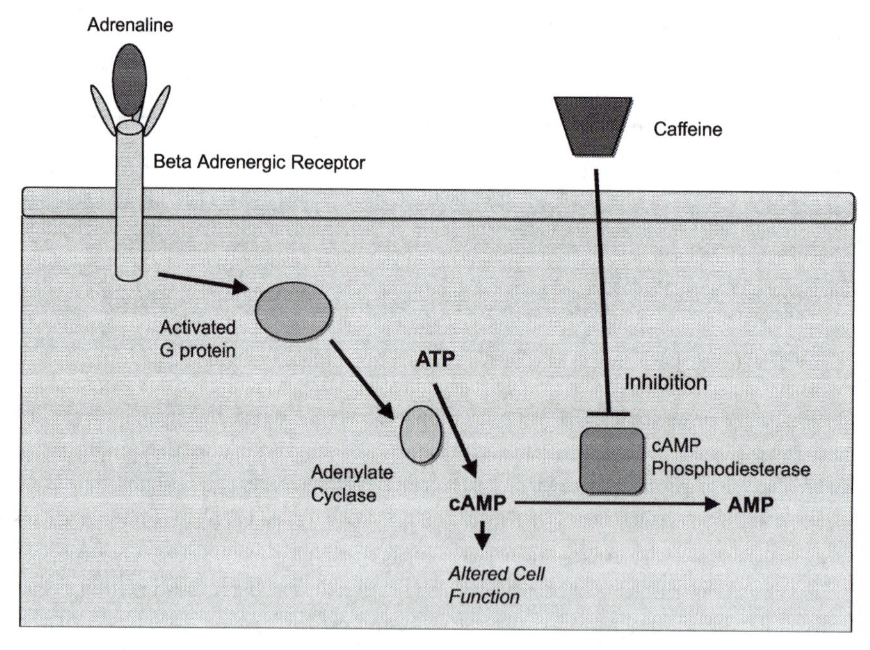

This figure shows the pathway for the production of cAMP beginning with the ß-adrenegic receptor. The effect of cAMP can be decreased by an enzyme that breaks down cAMP, but as you can see here caffiene can prolong the effects of cAMP by inhibiting its breakdown

signal molecule that is produced by the **adrenal gland.** Adrenaline can bind receptors on adipose tissue, which results in the breakdown of fats, yet it can also bind receptors on heart muscle cells and cause an increase in the rate and force of contraction.

Armed with information about signaling, Nate was prepared to find out how and where caffeine affects the process. In Nate's situation, he had loaded his body with caffeine and put himself through the stress of studying for finals early into the morning. Normally, his body would have consistently secreted a small amount of adrenaline from the adrenal glands, but because he was under stress, he produced more. There are two types of **adrenergic receptors** on the plasma membrane of cells that respond to adrenaline. They are named α and β. When adrenaline finds β receptors on cells, they invariably initiate the same signal transduction cascade. When adrenaline binds the β receptors, the cascade includes producing a molecule called cyclic adenosine monophosphate (**cAMP**), although this cascade does not have the same end result in every cell type. When adrenaline binds to β receptors on heart muscle cells, a rise in cAMP follows, which results in an increase in both the rate and strength of contraction. After cAMP has done its job, an enzyme breaks it apart, and the heart goes back to beating normally, provided adrenaline levels are also normal. Caffeine inhibits the enzyme that breaks cAMP apart, so the effect of adrenaline is more pronounced because the second messenger that carries the message stays around longer. Therefore, any signal molecule that uses cAMP as a second messenger will have a more pronounced response when caffeine is present because the caffeine will allow cAMP to stay at higher levels for a longer time.

When Nate was lying in bed having consumed four heavily caffeinated drinks, he felt his heart beating harder and faster than usual. He was not dreaming. Caffeine was in the cells of his heart, inhibiting the enzyme that breaks apart cAMP, and was causing a harder and faster heartbeat. This was the same feeling Nate might have had if he had been startled awake and realized that he had slept through both of his exams. The immediate intense fear would cause a burst of adrenaline secretion from his adrenal glands, which would result in an immediate increase in the strength and rate of heart contraction.

The β adrenergic receptors (adrenaline receptors) are stimulatory receptors that are of great medical interest. A class of drugs called β **blockers** that are **antagonists** of the adrenaline β receptors has been developed. Such drugs are used to treat patients who would benefit from reduced stimulation of the β adrenergic receptors. People who have **hypertension** are prime candidates for the use of β blockers. In these situations, people have much higher than normal blood pressure and by treating with a β adrenergic receptor antagonist, the patient's heart beats more slowly and with less force, which reduces the blood pressure. There are many other situations where β blockers may be of value, as for people who have experienced heart attacks or have irregular heartbeats.

Questions about this case:

1. If two different cells both have β receptors for acetylcholine, must they have the same physiological effect when acetylcholine binds the receptor?
2. Why would β blockers be valuable in treating people who have irregular heartbeats or have previously had a heart attack?
3. What is the difference between a growth factor, neurotransmitter, and a hormone?
4. What is transcription?

Questions to go deeper:

1. What is the primary second messenger for the α adrenergic receptor?
2. Besides inhibiting the enzyme that breaks down cAMP, what are some of the other effects of caffeine?
3. Acetylcholine has an immediate effect on heart cell physiology. Do you think the signal transduction pathway for acetylcholine in heart cells results in the gene transcription? Why or why not?
4. Not all receptors for signal molecules are confined to the plasma membrane. Give an example of one that is not.

An Asterisk in the Record Book

A Case Study of Performance-Enhancing Steroids

Until late in the summer of 2007 the Major League Baseball (MLB) record for career home runs was 755 set by Henry Aaron. Aaron was not unusually big for a professional baseball player, he stood 6 foot and weighed 180 pounds. Aaron retired in 1976. Similarly, until 1998, the record for home runs in a season was 61 and was set in 1961 by Roger Maris. Like Aaron, Maris was not a huge baseball player by today's standards. Maris's single season record fell in 1998 when Mark McGuire hit 70. That same year, Sammy Sosa hit 66. Only 3 years later, Barry Bonds hit 73 home runs and went on to break Aaron's mark for home runs in career.

What has happened to baseball? Maris's record that stood untouched and was considered almost unbreakable for 35 years has been broken twice in 4 years (not counting 3 seasons where Sammy Sosa hit over 61 home runs). Records that seemed impossible to break have fallen. What has changed?

For as long as sport has existed, the athlete has looked for a competitive edge. In the modern era, this often takes the form of performance enhancers. It is important to identify what performance enhancers are. Many years ago, they were just referred to as **steroids,** since steroids make up the most important category of chemicals that are now available. While most amateur and professional sports have a ban on steroid use, until 2002, MLB had no official policy. As a result, steroid use was fairly common and has landed Barry Bonds in the middle of a huge scandal. Are steroids the cause of the recent assault on the record books? To even suggest such a concept, we must understand what steroids are and what they can do.

In general, steroids are a group of chemicals with similar structures that are categorized as **lipids**. While molecules like **cholesterol**, which is involved as a structural molecule in cell membranes, are in this group, most steroids serve as **hormone** signals. Cells can send signals that are **endocrine**, **paracrine**, or **autocrine**. In addition, some cells can send signals that require cell-to-cell contact. Of these types of signals, steroids are most often sent in an endocrine manner. Endocrine signals are made in one **gland** or organ and sent through the bloodstream to all parts of the body. Any cell that has a **receptor** for the specific steroid in question will then respond to the signal. Steroids are **fat soluble** and cross the **plasma membrane** into the **cytoplasm** and **nucleus** easily. Steroid receptors are, therefore, not on the plasma membrane, like most receptors, but are in either the cytoplasm or nucleus. Steroid hormones alter cells by changing which **genes** are **transcribed** and, thus, which proteins are produced. There are many different steroid hormones, but the ones usually used for performance enhancement are the **androgen**-type **anabolic steroids**. Androgens are sex-specific male hormones, and hormones that are anabolic are those that promote growth and development. **Testosterone** is the best known of this type of hormone, which is made in the **testes** and **adrenal glands**. Responsive tissues have receptors for testosterone and respond to the hormone by **transcribing** genes involved in cell growth.

Testosterone has profound effects on the growth of muscle and bone. Muscle cells that come in contact with testosterone respond by producing more muscle proteins, which results in increased muscle mass. Each muscle is made of bundles of muscle cells, with each consisting of a large bundle of **fibrils,** along with the **nucleus** and **organelles** that we would expect to find in animal cells. The fibrils are a collection of thick **myosin** protein filaments and thin **actin** protein filaments arranged parallel to each other. Actin is oriented in a cage-like arrangement around the myosin in individual units called **sarcomeres**. Myosin has little appendages (called heads) that reach out and contact the surrounding actin cage. When muscles contract, the myosin heads attach to the actin and pull the ends of the sarcomere closer together. When all the sacromeres within a muscle are given the signal to contract (shorten) simultaneously, each fibril will shorten at the same time, causing the muscle to shorten in length. A muscle's strength is related to how many fibrils (rows of sarcomeres) pack each cell. As you might imagine, an athlete taking an anabolic steroid supplement over a period of time in conjunction with heavy workouts could become stronger and faster than he or she would be without the supplements. The steroids could, therefore, help an athlete be stronger, and stay stronger at an older age, than a comparable athlete who did not take the steroids.

Taking supplements like steroids does not come without **side effects**. Athletes are taking a form of the male sex hormone, so many of the masculine **secondary sexual**

characteristics will be enhanced. These would include an increase in anything from body hair to aggressiveness. One common side effect of taking steroid hormones is a suppressed immune system. As it turns out, some of the cells of the immune system express receptors for steroid hormones. Unlike muscle and bone, these cells die in response to hormones binding to the receptor. This is quite unusual. It wasn't long ago that scientists assumed that cells died only when injured. In the situation of steroid treatments, these WBCs respond to a normal hormonal signal by committing suicide. This normal death response is called **apoptosis**. White blood cells respond to high concentrations of steroids by going through apoptosis. Apoptosis progresses as the cell responds to the steroid signal by turning on enzymes that break up the cell's DNA and structural proteins. The cell ultimately shrivels without breaking open, and then is engulfed by a local **macrophage**. Apoptosis is a valuable response because these cells die and are eliminated without initiating the inflammation that would occur if the cell were to break open.

Why do some cells have the ability to die on command? There are a number of reasons. As limb buds grow in developing **embryos** of **vertebrates**, individual fingers and toes are formed not by excess growth of the cells that make the digits, but by the death (by apoptosis) of the cells between the digits. Cells may also go through apoptosis when they have had their DNA damaged so badly that it would be dangerous for them to try to replicate the DNA and go through cell division. Some virally infected cells will go though apoptosis to keep the **virus** from effectively replicating.

Scientists have also learned to use the apoptotic response of WBCs to steroids to their advantage. Treating with steroids suppresses the immune system. Therefore, any disease that is a result of an over-active immune system might be helped by steroid treatments. **Autoimmune diseases** and severe **allergies** both fall into this category, and people with these problems often find relief by steroid treatments. The key is to find a dose that suppresses the immune system enough to relieve the symptom of the disease without making the person susceptible to infections.

Despite the potential side effects and anti-steroid policies, athletes in a wide variety of sports are routinely caught using steroids to enhance their performance. With a newly established anti-steroid policy, MLB will try to curb steroid use among its players. But what about records that have fallen at the hands of players widely suspected of using performance enhancers? Many people argue that the records only be added to the books with an asterisk to indicate the circumstances surrounding them.

Questions about this case

1. What means are used to catch athletes using steroids?
2. Why do some people refer to apoptosis as "programmed cell death"?
3. What are some of the side effects of steroid use for performance enhancement not mentioned here?
4. What are some side effects of steroid use in disease treatment?

Questions to go deeper

1. Briefly describe how autoimmune diseases work.
2. What are some ways endurance athletes engage in performance enhancement by "blood doping"?
3. What are some other steroids and what effect do they have on cells that have the correct receptor?
4. Describe the action of myosin when muscles relax.

Reference

http://www.baseball-reference.com

Going Blind on the River

A Case Study of Vision and Visual Impairment

If you need to see a physician in Zapallo Grande, you can't make an appointment. To see a physician in this village in the Western jungle of Ecuador, you have to wait for the physician to make the appointment to visit the small, 30-year-old, one-room clinic. In the clinic, one doctor does everything from pulling teeth to performing abdominal surgeries to dispensing medicine. Disease is rampant in this part of the world, and people do not go to the clinic unless things are serious. When the local villages hear that the doctor will be visiting, the word is spread, so on the morning the doctor arrives, there is already a line. After Dr. Ron and I got off the small, single-engine airplane, I followed him through the crowd. A woman blocked our way and handed Dr. Ron a rooster tied up at the feet. He took it and in turn handed it to me, explaining that he had saved her husband's life on one of these visits and we would be eating rooster soup for dinner. As we got to the clinic, Dr. Ron surveyed the line. About 25% of the people were there to get teeth pulled. They had not learned to brush their teeth so they had abscesses that swelled one cheek the size of half a cantaloupe. Another 25% clearly had **malaria** and had been brought by a family member to get treated and (hopefully) recover in the hospital hut next to the clinic. Another 25% of the patients had a wide variety of ailments that only a physician trained in tropical medicine could diagnose. What surprised me, though, was the last 25% of the patients in line. Dr. Ron knew these patients were coming because they are there every time he visited. They did not seem to have anything physically wrong with them; they showed no obvious symptoms. When I asked Dr. Ron about them, he explained that these patients had **onchocerciasis**, which is also known as **river blindness**. "Oncho what?" was my reply. I had never heard of such a thing.

Over the next couple of days, working at the clinic, Dr. Ron explained in detail about the parasite ***Onchocerca volvulus*** and how it can take away a person's sight.

Onchocerca volvulus is a parasitic worm that in its adult form lives in **nodules** in the skin of the human host. Male and female worms are found paired in the nodules reproducing and releasing **larvae** that travel throughout the body in the bloodstream. When a **black fly** bites the infected human, the larvae can be taken up by the fly. Inside the fly, the larva goes through a couple of developmental stages and then can be deposited into another human host during the next bite. The fly is required for the life cycle of the worm to continue. Since the fly only reproduces in rivers, the disease is always found near a river, thus the name "river blindness." Once the juvenile enters the human, it matures into an adult and takes up residence in a **subcutaneous** nodule. Once I took a closer look at the patients that had come to the clinic with this disease, it was clear that they all had nodules. But, the nodules alone do not explain why the disease is so devastating. To understand the disease, we first must understand the eye.

The eye is a quite simple and effective imaging system. Light enters the eye by crossing a protective covering called the **cornea** and enters the **lens** by passing through an opening called the **pupil**. The size of this opening is regulated by the **iris**. The iris is a donut-shaped piece of tissue that gives the eye its color. In low-light conditions, the iris opens to collect more light, and in bright light, it closes to restrict the amount of light that enters. The opening and closing of the iris is involuntary and will occur any time the brain senses light (or the lack of light) coming through the iris. The function of the iris is a common diagnostic tool used to check for brain injury. If light comes through the iris and the brain can sense it and is able to respond,

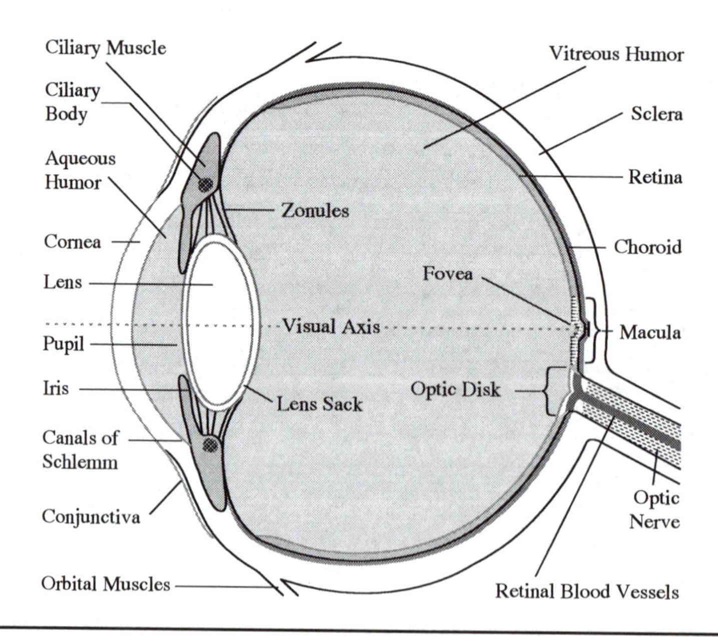

then the iris should constrict involuntarily, implying the brain has not been damaged. I am sure you have seen a doctor or paramedic on a TV show do this by shining a flashlight in a patient's eye and then saying something like, "The patient's pupils are responsive." When incoming light crosses the lens, it is focused so that a sharp image is displayed on the back of the eye, called the **retina.** The lens changes its focus by the contraction and relaxation of muscles that attach to its edges. The nearer the light source the more the light needs to be bent to focus on the retina. Light from a source nearby would require a thicker lens and the muscles controlling the lens would accommodate by contracting. On the other hand, light from a source farther away requires less bending to focus it on the retina, and the eye responds by the muscles relaxing. Embedded in the retina are **photoreceptor cells** called **rods** and **cones.** These cells have light-sensitive ends that sit in the retina and connections to the **optic nerve** on the other end called **synapses.** Rods contain a light-sensitive molecule called **rhodopsin** and cones have one called **photosin.** Rhodopsin is a more sensitive molecule that transmits black-and-white images in low light, while photosin transmits colored images, but is less sensitive. Cones transmit colored images by existing in three varieties that respond to the light of different colors of blue, red, and green. The pattern of stimulation of the different cones results in the ability of the human to differentiate a vast array of colors. These rods and cones are found at a particularly high density at the center of the visual field, called the **fovea,** and are nearly absent at the spot where the optic nerve connects to the back of the eye, called the **blind spot.** To send an image to the brain, the rhodopsin and photosin in the rods and cones respond when light hits the retina by sending a chemical message through the photoreceptor cells to the point where the cells interact with neurons. The rods and cones then send a **neurotransmitter** across the synapse to the adjacent neuron. When the neurotransmitter reaches the adjacent neuron and binds to a **receptor**, the neuron converts the message to an electrical charge, which it sends on to the brain.

In the case of the patients who came to Dr. Ron's clinic with onchocerciasis, each had adult worms living in subcutaneous nodules. As these adult worms produced larvae, many of the larvae ended up in the blood vessels that feed the photoreceptor cells of the retina. In this location, the immune system is able to use **WBCs** of the **cell-mediated immune system** like **eosinophils** to kill some of the larva. As some of the larva die, more cells of the immune system are called in to help with the problem of the dying and dead larvae. Unfortunately, as these immune cells collect in the retina they cause irreparable damage, or scarring. The very immune system designed to protect you from these foreign invaders is what is causing people with onchocerciasis to go blind. The treatment is pretty simple, but not fast. Each patient is given a drug that effectively kills the juveniles, but then each nodule is identified on every patient and surgically removed. It's hard to imagine doing surgery in 100-degree heat with near 100% humidity in conditions that are less than sterile. There is a risk of infection, but it beats the risk of going blind.

You probably were not familiar with river blindness until you read this case, but the explanation of the eye structure and function holds the answers to a couple of medical problems you undoubtedly are familiar with. People who are **nearsighted** and **farsighted** lack the ability to correctly focus some images onto the retina. When a person is nearsighted, also called **myopia**, light from a distant source cannot be focused because the lens cannot be flattened enough. As a result, a myopic or nearsighted person can see well at short distances, but not at a distance. The opposite is true with people who have farsightedness, which is also called **hyperopia**. The farsighted person cannot make their lens thick enough to focus near images, but do not have a problem focusing on objects at a distance. Hyperopia often takes the form of **presbyopia**, which is a word that takes its root from the Greek meaning "old man." Ultimately, farsightedness often develops as people get older, and is the reason many people need "reading glasses" or "bifocals" as they get older. These impairments can be diagnosed with a **visual acuity test**, which you probably know as the eye chart with the big "E" on the top. The chart is to be read at a distance of 20 feet, and if you can read the lines properly at 20 feet, then your vision is said to be 20/20. If you read the chart at 20 feet in the same way a person with 20/20 vision can read the chart at 100 feet, then you would have 20/100 vision. Usually each eye is tested separately since they often have different acuities. Corrective lenses are available for either of these impairments and are custom made according to the specific acuity of each eye.

We tend to take our vision for granted. People in places where river blindness is rampant do not. Their sight provides their only means of work and the ability to gather food. It has been said that going blind on one of these rivers is a death sentence.

Questions about this case

1. Why is night vision mostly black and white?

2. How does the eye of a hawk differ from that of a human eye? How does the eye of an owl differ from the human eye?

3. On the 20/20 scale what is considered legally blind? What does 20/100 acuity mean?

4. What are "bifocals"?

Questions to go deeper

1. What is glaucoma and how does it cause blindness? How is it treated?

2. What is astigmatism?

He Is Two, Shouldn't He Be Walking by Now?

A Case Study of Muscles and Muscular Dystrophy

At 2 years, Joshua was not walking. He tried to get to his feet and it seemed he could balance, but he would simply fall back down. His parents, Tim and Stephanie, were concerned. They had seen other kids walk at 12 months, but most people they consulted with gave the same answer. It just takes longer for some kids to walk, talk, read, or whatever. But, as they watched carefully as Joshua tried again and again to get up, they could tell something was wrong. He simply did not seem strong enough. Tim and Stephanie decided to take Joshua to the doctor. The pediatrician felt Joshua's legs and arms. He put Joshua on the floor and, as if it were scripted, Joshua tried to get to his feet and walk, but fell back to the floor. The pediatrician asked that Tim and Stephanie allow some blood to be drawn for a blood test. The blood was drawn, and Tim and Stephanie took Joshua home. Within a week, they received a call from the doctor requesting a follow-up visit. When they arrived for the appointment instead of being led into an exam room, the nurse led them into the doctor's office and they immediately sensed something was wrong. The pediatrician explained that Joshua had an enzyme in his blood that should not be there. The enzyme, creatine phosphokinase, is usually found at high levels only inside muscle cells. Finding it in the blood indicates some type of muscle damage. The pediatrician continued by explaining that Joshua's apparent muscle weakness and the blood test together suggested that Joshua may have **muscular dystrophy** (MD), but a muscle **biopsy** would be needed to confirm the diagnosis.

Tim and Stephanie were stunned. Of course, they had heard of MD, but they had no idea of the ramifications of the news. After the biopsy, they took Joshua home and over the following week, they used the Internet to learn about MD. Before they could even understand how MD caused problems with muscles, it was important to understand how muscles normally work. They discovered that there are three types of muscles: **skeletal**, **cardiac**, and **smooth**. Smooth muscle is found in organs like the intestines and uterus, and cardiac muscle is found only in the heart. Skeletal muscle, on the other hand, makes up the **voluntary muscles** associated with the skeleton. Each type of muscle is made of bundles of muscle cells that are really individual muscle fibers. Each cell (or fiber) consists of a large bundle of **fibrils,** along with the **nucleus** and **organelles** that we would expect to find in animal cells. The fibrils are a collection of thick **myosin** filaments and thin **actin** filaments arranged parallel to each other. The diagram below shows how the actin is oriented in a cage-like arrangement around the myosin in individual units called **sarcomeres.** Myosin has little appendages (called heads) that reach out and contact the surrounding actin cage. When the signal to contract is given, the myosin heads attach to the actin and change orientation by flipping toward the center of the sarcomere. This action requires **ATP** as it pulls the actin cages together, effectively shortening the sarcomere. The end of each sarcomere, called the **Z-line,** is also the end of the adjacent sarcomeres. So, if all the sacromeres are given the signal to contract (shorten) simultaneously, then each fibril will shorten at the same time, causing the muscle to shorten in length. To relax, the myosin heads simply release the actin filaments and the sarcomere springs back to its original length. This model of muscle contraction is referred to as the **sliding-filament model.**

In the sliding-filament model it is clear that a muscle only uses ATP when it contracts. Therefore, a skeletal muscle can move a joint forcefully only in one direction. The ends of skeletal muscles are connected to bones with tough fibrous **tendons.** For example, the muscle in the front of the upper arm is called the **biceps bracii** and it connects to the top of the bone of the upper arm called the **humerous** at one end and

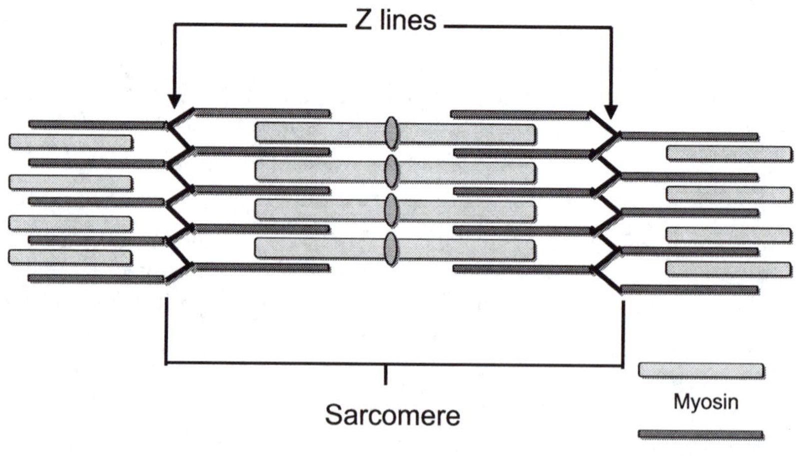

Diagram of the structure of a sarcomere. The myosin filaments are arranged between parallel actin filaments and the Z-lines are brought closer together when myosin appendages (not shown) temporarily attach to the actin and pull them towards the center. Muscles contract when rows of sarcomeres shorten simultaneously.

the **radius** bone in the forearm at the other. When the biceps bracii contracts, it draws the hand closer to the shoulder. This **flexing** motion would allow a person to do something like touch his or her nose. To return the hand to the side requires no ATP; the person just relaxes and the hand falls due to gravity. But, what if a person wants to forcefully push their hand back to their side? The biceps cannot move the forearm with force in that direction. In this situation, a joint needs an **antagonistic pair** of muscles, one to forcefully move the joint in each direction. In the human arm, the **triceps bracii** muscle found in the back of the upper arm is the other member of this pair. It connects to the humerous in the upper arm and the **ulna** bone in the forearm, and contraction of the triceps muscle allows a person to **extend** the arm with force. Antagonistic pairs of muscles are found wherever joints need to forcefully move in more than one direction. Other pairs are **latissimus dorsi/deltoids** in the shoulder and the **quadriceps/hamstrings** groups in the leg.

With this understanding of muscles under their belt, Tim and Stephanie began to try to discover what they could about MD. They found that there are many types of MD with different muscle groups involved, symptoms, and severities. From Joshua's symptoms, it would be most likely that he would have the most common form of MD called **Duchenne muscular dystrophy**. While all the types of MD are due to genetic abnormalities, Duchenne MD is a result of a mutation in the **dystrophin gene** found on the **X-chromosome**. The disease is considered **X-linked recessive**. It is recessive since people can get the disease only if both copies of the gene are lost to mutation. What do we mean by both copies? Remember that humans have 46 chromosomes that exist in 23 pairs. This gene is found on the sex-determining chromosomes, which is the 23rd pair. A woman by definition would have two X chromosomes and a man an X and a Y chromosome for this pair. For a woman to get Duchenne MD, she would have to inherit two mutated X chromosomes—one from her father and one from her mother. Since a male only has one X chromosome he only has to inherit one mutated X chromosome. That mutated X must have come from his mother because a boy must have one Y chromosome and that could only come from his father. For this reason, we find that Duchenne MD, like all X-linked traits, is most often inherited by boys from their mothers.

Still, Tim and Stephanie's burning question was, "What goes wrong in MD to cause catastrophic muscle weakness?" The answer lies in the function of the product of the dystrophin gene. The dystrophin protein seems to serve a slightly different function in muscles than it does in non-muscle tissues, and its exact function in muscles is still being debated. From our best understanding, dystrophin is a structural protein that helps link the long, repeating actin cages to adjacent cells. When all the fibrils in one cell contract and that cell shortens, it pulls on the surrounding cells. Providing all these cells stay attached to each other and the actin cages remain attached to each other, the muscle as a whole shortens. In Duchenne MD, it appears that the actin cages do not attach well to the outside of the cells and the cells do not stick together well. When the signal for contraction comes, the fibrils contract, but the internal actin connections and cell-to-cell connections are lost. The fibrils shorten, but the overall muscle does not.

As Tim and Stephanie continued their research, the news got even worse. Duchenne MD is progressive. If Joshua did have Duchenne MD, then the muscle weakness that began in the legs and hips would likely progress to the **diaphragm** and the heart. Breathing would ultimately become difficult, and heart failure would be eminent. Most boys with Duchenne MD do not reach 25 years of age. With such a terrible **prognosis** for boys with Duchenne MD, Tim and Stephanie did not sleep much over the next week. But, they were ready with questions when the doctor called, no matter which way the diagnosis went.

Questions about this case

1. Shivering is a type of muscle contraction that is a response to being cold. How does shivering differ from regular contraction? Why does the body respond to cold in this way?

2. When exercising, why is it important to work joints by both flexing and extending?

3. What would a mother be called who had one mutated X chromosome for Duchenne MD?

4. Women can contract most X-linked diseases, they just don't nearly as often as men. Why do women almost never get Duchenne MD?

Questions to go deeper

1. What role does the endoplasmic reticulum have in muscle cells that is critical for contraction?

2. What is the signal that tells muscle cells to contract? Where does the signal originate?

3. How does Duchenne MD differ from Becker's MD? Why would Tim and Stephanie be pretty confident that Joshua did not have Becker's MD?

4. What is the relationship between the dytrophin gene and dystrophin protein?

A Massive Heart Attack at 26

A Case Study of LDLs and Heart Disease

This illustration shows the cut-away view of an artery with the artery wall shown in black, the healthy endothelial cells in dark gray, and the artery-clogging plaque in light gray.

Fire department paramedic units are called to patients having **heart attacks** regularly, so these calls rarely surprise them. This call would be different. As the paramedics hurried into the house, they were surprised to find their patient was a 26-year-old male, named Steve. He was in full cardiac arrest, and it did not look like they could save him. As the paramedic crew worked feverishly to save Steve, some of the fire crew comforted Steve's frantic wife, Melanie, and worked on getting some family history. It's not often that a 26 year old has a massive heart attack, but the reason Steve was in this situation was the real surprise.

Melanie explained that Steve suffered from a rare genetic disease called **familial hypercholesterolemia** (FH), which causes rapidly accelerated **atherosclerosis**. Melanie went on to explain that this was not Steve's first heart attack, and Steve's doctors had warned them this next heart attack might be fatal. The crew continued to work on Steve and actually got him stabilized enough to transport him. The ambulance arrived, and Steve was quickly loaded up and taken to the hospital. As the ambulance left, the fire and paramedic crew discussed what they had witnessed. Paramedics roll to literally dozens of heart attacks a year, but 26-year-old victims are rare. Even rarer are patients with FH. For most paramedics, this is something they only read about in textbooks.

To understand FH, first we must understand the term. "Familial" means that the disease runs in the family; in other words, the disease has a genetic basis. The word "hypercholesterolemia" has three parts: the prefix *hyper* means to have an excess of, *cholesterol* is a fat-soluble molecule that is used in making cell membranes, and the suffix *emia* is related to the blood. So hypercholesterolemia is simply high blood cholesterol levels. What is so bad about that? Scientists have discovered that there is a direct correlation between high cholesterol levels and atherosclerosis. Atherosclerosis used to be called "hardening of the **arteries,**" but what is actually happening is arteries collect cholesterol-containing deposits called **plaques**, and as the deposits grow, they block blood flow through that artery (the plaques make that area of the artery hard). Unfortunately, the most common arteries involved in atherosclerosis are the **coronary arteries** that supply the heart muscle with **oxygenated blood**. Without adequate blood supply, this tissue will die, which is what happens in a heart attack.

People with FH have a genetic defect, or **mutation**, in a single **gene**. This gene provides the **code** to produce a **protein** called the low density lipoprotein receptor (**LDL receptor**). Humans have 46 **chromosomes** that are arranged in 23 pairs. Each member of a pair codes for the same traits, although they will carry different information. For example, the genes that code for blood type are found on chromosome 9. You have two chromosome 9s and since you inherited one from your mother and one from your father, they could carry the coding for different blood types. The LDL receptor gene is found on chromosome 21, and each of the pair of your chromosome 21s should have the same LDL receptor gene since this gene does not normally differ from person to person. In Steve's situation he has inherited two chromosome 21s with a damaged copy of the LDL receptor. He, therefore, does not make any functional LDL receptor.

The key question here is, what does the LDL receptor do? To make any protein (like the LDL receptor), a gene must be copied into a messenger RNA (**mRNA**) in the nucleus, during a process called **transcription**. The mRNA is then sent to the **cytoplasm** of the cell, where the mRNA code is read by a **ribosome** and a protein is assembled with the exact amino acid sequence, as determined by the original gene. This process is called **translation**. Proteins like the LDL receptor that end up on the cell surface are translated by ribosomes that are on the surface of the **endoplasmic reticulum** (ER). These proteins are placed into the membrane that surrounds the ER and from there, they become a part of **vesicles** that pinch off the ER and are sent to the **golgi apparatus**. The golgi then sorts the different proteins and packages them into vesicles so that all the proteins that need to go to a common location are together in one vesicle. The LDL receptors are packaged as part of the membrane of vesicles that are shipped to the **plasma membrane**. When the vesicles fuse with the plasma membrane, the LDL receptors become proteins on the cell surface.

Once the LDL receptor is on the cell surface, it can do its job. In tissues like **liver,** the LDL receptor is placed on the surface of cells that line blood vessels. As **LDLs** stream by, they bind to the receptor. These packages of proteins, lipids, and cholesterol bind to their receptors, which are collected together into a vesicle that will bring the LDLs into the cell. This is called **endocytosis**. The incoming vesicle is ultimately fused with an organelle called a **lysosome**. The lysosomes contain powerful digestive **enzymes** that break apart the components of the LDLs so they can be used by the cell. The enzymes of the lysosome are produced in a similar manner as the LDL receptor. They are also translated by ribosomes on the ER, but are passed to the inside of the ER instead of being stuck in the membrane. As transport vesicles form, these proteins are sent to the golgi, and then they are collected into vesicles and sent to the lysosomes. When the vesicles fuse with the lysosome, the contents of the vesicles are released inside the lysosome.

It's easy to see why people with FH have heart attacks at such an early age. People constantly make LDLs that are placed into their bloodstream. If no LDL receptors are made, then the LDLs rise to incredibly high levels with no means of taking them out of the blood into cells. LDLs are absorbed into the plaques that lead to atherosclerosis. Therefore, high levels of LDLs lead to rapidly progressing atherosclerosis. The result is a heart attack that occurs at an abnormally early age. A simple test can determine cholesterol levels in the blood. Normal blood cholesterol is 200 mg/dl or less. However, this number represents two different components: LDLs and HDLs. We

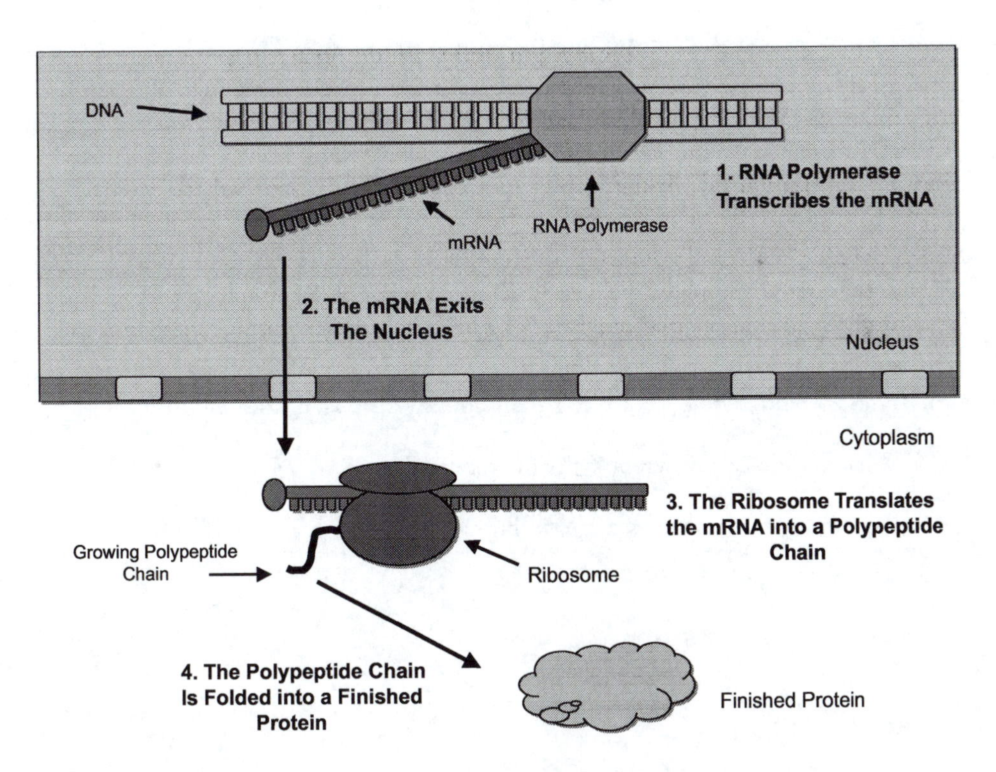

This figure illustrates the steps required to produce a protein from the DNA code.

have already discussed LDLs. These are sometimes called "bad cholesterol" because of their link to heart disease. It is good to keep LDL levels below 130 mg/dl. HDL (high density lipoproteins) are also collections of proteins, lipids, and cholesterol, but carry much less cholesterol and lipid. They are often called "good cholesterol" because higher levels of HDLs are associated with a reduced risk of heart disease. It is healthy to have HDL levels of above 40 mg/dl. A person with FH can have total cholesterol levels of greater that 450 mg/dl!

Even if Steve survives this heart attack, the **prognosis** does not look good. The drugs that are commonly used to reduce LDL levels in the blood have little effect in patients with FH. Yet, Steve and Melanie know the best they can do is to continue to choose healthy habits and submit to aggressive medical efforts.

Questions about this case

1. What happens to the LDL receptors after they make it to the lysosome?
2. What is an angiogram and how might it be used as a diagnostic tool for someone who suffered a heart attack?
3. Where are proteins made and sent?
4. What else does the ER do besides being the location for translation?

Questions to go deeper

1. What might happen if a person inherited only one mutated LDL receptor gene?
2. Why would Melanie know with such certainty that Steve had FH?
3. Normal blood cholesterol is less than 200 mg/dl. What does mg/dl mean?
4. What are the two different types of secretion?
5. How many proteins does a typical human mRNA carry the code for?

Thank Goodness, Dad Doesn't Have Cancer

A Case Study of Liver and Pancreas Function

I got a phone call late one evening that my father had driven himself to the hospital because he was experiencing severe abdominal pain. I arrived at his room shortly after he was admitted, with the typical set of unanswerable questions. Since I know something about biology, I wanted some answers beyond "we are doing some tests." The attending physician came by his room later that night and explained that he simply did not know what was causing the pain. While that is understandable, I still wanted to know what the most likely candidates for the cause of the pain were and what tests he was running to find the answer. When a physician is confronted with a patient with an unknown ailment, a **differential diagnosis** is used to help determine what the problem is. This means that the physician will use a complete **medical history** and the list of **symptoms** to predict what the most likely candidates for the ailment are. My father was in his middle 60s, with no previous personal or family history of abdominal pain. Upon examination in the emergency room, it was determined that the location of the pain was in the area of his **liver**, **gall bladder**, and **pancreas**. This ruled out **ulcers** as the source of the pain, which is a common source of abdominal pain in older men. Two common sources of liver damage in men of my dad's age are alcohol abuse and **hepatitis**. Since dad doesn't drink and had not previously been diagnosed with hepatitis, these were unlikely.

I went by the hospital the next day and visited with my father. He said the physician had been by to see him and the additional tests had indicated that his pancreas was the most likely cause of the problem. I knew immediately that was not the best news. The two most common sources of pancreas-related pain for men like my father are **pancreatitis** and cancer. Pancreatitis, which is inflammation of the pancreas, is very painful and could be the result of a number of other problems with a myriad of **prognoses**. On the other hand, I knew pancreatic cancer is also very painful and has a high mortality rate.

Why are these types of problems with the pancreas so painful? The answer is found in what the pancreas does. The pancreas is a **gland** that is found in the center of the abdominal cavity below the liver and **stomach** and encircled by the **small intestine**. Its job is to produce two types of molecules for **secretion**. The first type is hormones, including **insulin** and **glucagon**, which regulate blood sugar levels. These hormones are released in the bloodstream to affect cells throughout the body. This type of secretion is called **endocrine** secretion. The inappropriate production of these hormones and the body's response to them form the basis for the diseases **diabetes** and **hypoglycemia**. Only a very small percentage of pancreatic cells are responsible for hormone production, and these hormones are not the source of intense pain from pancreatic diseases. The second type of molecules produced by the pancreas is **digestive enzymes**. Enzymes to break apart **fats**, **proteins**, **nucleic acids**, and **carbohydrates** are produced in the pancreas and released into a **duct** that carries the enzymes to the first section of the small intestine, called the **duodenum**. The enzymes help break down food so that the small intestine can absorb the food molecules. This type of secretion is called **exocrine** secretion. When the pancreas becomes inflamed or damaged, the digestive enzymes leak into surrounding tissues and cause damage, resulting in pain.

My father was ultimately diagnosed with pancreatitis of unknown origin. It was resolved over a 3-day stay in the hospital and has not returned. That is the best possible diagnosis and outcome for someone with his symptoms. But, what if this whole story played itself out in some other part of the world, like China? Is there any difference between my father in Southern California and a man in China in his 60s with severe abdominal pain when they visit a physician? From the physician's standpoint, of course there is. Had my father been living in China at the time of his illness, the physician would employ the same principles of differential diagnosis, but the list of likely causes of abdominal pain would change. He might get asked questions like, "Do you spend time wading or swimming in the rivers?" or "Do you eat uncooked fish?". Living in China would expose my father to diseases that could cause pain localized to the liver, gall bladder, and pancreas that he would not come into contact with in the United States. For example, grass carp is a delicacy in China that is often eaten raw. Carp are farmed in large ponds and they eat vegetation. Ponds are often fertilized to encourage plant growth. Unfortunately, grass carp carry a human parasite called ***Clonorchis sinensis***. The parasite develops in aquatic snails and infects the carp by penetrating the skin in a juvenile form. The juvenile grows in the muscle of the carp. When humans eat the raw carp the juvenile is released and grows into an adult worm in the intestinal tract. The adult begins to lay eggs that are released in the human feces. Since the snail must come into contact with eggs that are shed in human feces to complete the cycle, good sanitation could eliminate the problem. This answer may sound simple, but it is not easy. Rural China is lacking in infrastructure and finances to make sanitary handling of human wastes a reality. Moreover, human feces is the cheapest possible fertilizer, so it is intentionally spread in carp ponds to promote vegetation growth.

Ultimately, the adult worm often finds its way to the **bile duct** and can work its way up the duct to the gall bladder and liver. The bile duct connects the gall bladder to the small intestines. Like the pancreas, the liver is a gland that makes molecules for exocrine secretion. These molecules are stored as **bile** in the gall bladder on the underside of the liver. The bile duct connects the gall bladder to the duodenum so the contents of the gall bladder can be regularly released into the intestines. The liver produces a variety of molecules for both endocrine and exocrine secretion, and molecules to assist in digestion. For example, it produces **bile salts**, which help break up fats and make them easier to absorb. The liver also produces a large amount of waste molecules. It receives the blood that is returning to the heart from the intestines. As a result, the liver is the first to receive freshly absorbed food molecules and toxins, alike. By chemically altering toxic molecules, the liver makes them less toxic and then secretes them with the bile so they can be eliminated in the feces. An example of this process is the detoxification of alcohol. The liver uses an enzyme called alcohol dehydrogenase to make alcohol less toxic, but in doing so, it is not immune to the toxic effect of alcohol. Alcohol causes cell damage and inflammation, which will result in replacing normal functioning liver cells with scar tissue in a process called **cirrhosis.** In addition, the liver controls the production of nitrogen-containing waste molecules. All animals need a means of releasing nitrogen-containing waste that is a result of the breakdown of proteins and nucleic acids. In humans, the liver forms urea that is then collected and released by the kidney. The liver also controls blood glucose levels by responding to insulin and glucagon. It responds to insulin by absorbing glucose from the blood and storing it as glycogen. Conversely, in the presence of glucagon, the liver breaks down glycogen and releases glucose. Finally, the liver also produces blood **plasma** proteins that are involved in lipid transportation and blood clotting.

It should be obvious from the diverse functions of the liver that compromising the function of this organ can affect many other systems. *Clonorchis* infections can result in a clogged bile duct, a clogged pancreatic duct, and/or loss of liver function, which could result in a number of symptoms of which pain is just one. Things may have been very different had my father spent 3 months in China prior to driving himself to the emergency room. Without the knowledge that dad had traveled to China, diseases like a *Clonorchis* infection would be at the very bottom of a long list of possible causes of abdominal pain for a physician in the United States.

Questions about this case

1. How would a doctor most easily diagnose a *Clonorchis* infection?
2. What is cirrhosis? What causes it?
3. What are diabetes and hypoglycemia, and how do they relate to liver function?
4. People often talk about how the liver "purifies" the blood. According to the case, what is really happening?

Questions to go deeper

1. What are some other diseases that are prevalent in parts of the world that do not have good sanitation?
2. What does "inflammation" mean?
3. What is the difference between endocrine and exocrine secretion?

"I Can't Have Lung Cancer; I Don't Smoke"

A Case Study of Smoking and Lung Function

Jack was in great shape for a guy in his 60s. He ate well and stayed active playing tennis. He had scheduled this visit to the doctor for a routine checkup; he just wanted to make sure he stayed healthy. As the doctor asked questions about his health, Jack mentioned that he had a persistent cough. He had thought nothing of it and only mentioned it when the doctor asked. His doctor didn't have any noticeable response to Jack's comment. The doctor followed up by asking if Jack ever smoked, but he never had, so the conversation seemingly ended. The doctor wrote out a prescription for a refill on Jack's medication and a request for routine blood work. Jack was a little surprised when his doctor also wrote a referral to the radiologist for a chest x-ray.

"Why do I need the x-ray?" Jack asked.

"Well, you really don't need it," Dr. Malcom responded. "It's just precautionary."

"What are you concerned about?" Jack replied.

"Lots of people have persistent coughs for any number of reasons. But as a 60-year-old male the cough could be the early signs of cancer," explained Dr. Malcom. "Jack, you don't smoke and you never have, so I really don't think you have cancer. A chest x-ray is a pretty easy way to make sure you don't, and early diagnosis for **lung cancer** is the key to successful treatment."

Jack was satisfied with the explanation and left the office without much concern. He made the appointment for the chest x-ray in the following week and was told that results would take 3 to 5 days. Dr. Malcom's office called Jack a couple of days later to make an appointment to discuss the results. Jack was immediately concerned. When he asked the receptionist what the results of the x-ray were, she said that Dr. Malcom would discuss that with him. They made an appointment for the following day and, needless to say, Jack did not sleep well that night. "It must be serious," he thought, "the receptionist would not even tell me the results, and how else could I get a next-day appointment at the doctor?"

As Jack walked into Dr. Malcom's office he could tell something was wrong. Dr. Malcom explained that the x-ray showed a dense spot the size of a golf ball in his left lung, which looked like cancer. Jack was stunned! "But, I don't smoke," Jack responded.

Dr. Malcom continued, "Yes I know, but people who don't smoke sometimes get lung cancer, too. This is the exact reason we did the x-ray to begin with, Jack. By what I can tell from the x-ray, I think we caught it early."

Within a week, Jack was in surgery, where half of his left lung was removed. The pathology report indicated that Jack had an aggressive small cell **carcinoma**, but it had not **metastasized**. In further consultations with the **oncologist**, Jack was told that there was no question that the whole tumor was removed, and further treatments would not be necessary. Jack doesn't know how lucky he is. Five-year survival for lung cancer patients with a tumor that has not metastasized is greater than 90%, but it is less than 20% if the tumor has metastasized. Dr. Malcom was correct, early diagnosis is the key to successful treatment for lung cancer.

Jack had an intriguing question: How could he get lung cancer if he does not smoke? Cigarette smoke contains many chemicals that have been shown to help cause cancer, and there is no question that smokers have a much greater chance of getting cancer than non-smokers. The chemicals in cigarette smoke that help cause cancer do so by causing damage to **DNA**. DNA is our genetic material, and it is organized in a series of **genes**. Each gene provides the information on how and when to make a specific **protein**. Damage to DNA can result in damaging this information. If the damage prevents a cell from making a protein that helps tell the cell when to grow, then the ability to control cell growth is altered. Cancer is usually the result of more than one event of DNA damage or **mutations.** A collection of mutations within a single cell that cause that cell to grow out of control would be considered cancer. The problem is that many chemicals can cause DNA damage. In fact, there are so many that there is no reasonable way to avoid them all. Even foods we eat every day, such as peanut butter and mushrooms, contain small amounts of chemicals that can help cause cancer. In addition, some people inherit mutations that make it more likely that they will contract cancer. There is no way to avoid every chemical that could cause cancer, so even people who don't smoke can get it.

It is clear that cigarette smoking greatly increases the risk of cancer because cigarette smoke contains very high levels of many different **carcinogens**. The good news is if a smoker quits for 15 years, the risk of cancer is reduced back down to the level of a non-smoker. Cigarette smoking causes more problems than cancer, and some of them are due to damage that cannot be reversed even if the smoker quits. **Emphysema** is a good example. In emphysema, the chemicals in cigarette smoke cause damage to the lung tissue. When you breathe air in, it travels through the **larynx** and down the **trachea,** which splits into two **bronchi**. Each bronchus enters one lung and progressively branches into smaller and smaller passages, until a respiratory **bronchiole** is reached. This bronchiole is the last tubular passageway, and it opens into a large collection of sac-like openings called **alveoli**. The alveolar sacks and the bronchiole look something like a bunch of grapes. Your lungs have in excess of 500 million alveolar sacs, which are very fragile and are lined with blood vessels. As emphysema progresses, the chemical in cigarette smoke irreparably destroys the walls of the alveoli.

Why is alveoli destruction a problem? From the description above, you can see that your lungs are more than big bags that inflate with air. The goal of the lungs is gas exchange, meaning, exchanging the carbon dioxide in the blood with the oxygen that is breathed in. Because these gases simply diffuse in order to move from place to place, it is important that the distance the gases travel be short and the surface area of the membrane they diffuse across be large. What does this mean? For oxygen to travel out of the lung into the blood by random diffusion, it is important that every oxygen molecule does not have to travel far to find the membrane that lines the alveoli (which have blood vessels surrounding them). Lung anatomy is well suited to accommodate for this. The branching system of the lungs makes the final alveoli so small that the distance to the nearest blood vessel is very short. If the alveolar sacs are destroyed, then when air is breathed in, there is a diminished ability to have it diffuse into the blood.

Emphysema takes many years to progress to a serious stage. It starts as cigarette smoke paralyzes the small hairs called **cilia** that line the branching system of bronchi. Cilia are responsible for catching foreign particles in the **mucus** in which they are covered. Then, using a beating motion, they move the mucus and trapped foreign particles up to the esophagus, where they are usually swallowed. Those of you who are heavy smokers or who live with one are familiar with "**smoker's hack**." Over night, as a smoker sleeps, the mucus accumulates in the lungs, causing difficulty in breathing in the morning. Heavy smokers often go through a routine of severe coughing to rid their lungs of the accumulated mucus. With the cilia not functioning, bacteria and particulate matter from smoke accumulate in the lungs and cause irritation. The smoke also damages cells of the immune system so the cells designed to help kill bacteria and remove foreign particles cannot help, causing the problem to worsen. Ultimately, alveoli are destroyed. People with advanced emphysema have had so much alveolar damage that even though they breathe in air, they suffocate from the lack of gas exchange. Often they are given pure oxygen to breath to help them get as much oxygen in their blood as possible. There is no cure for emphysema; once lung tissue is destroyed, it cannot be repaired.

A common symptom of progressing emphysema is difficulty exhaling. To inhale, you contract the muscle below your lungs called the **diaphragm**. By contracting your diaphragm, you increase the volume of your lungs, which reduces the air pressure in your lungs below that of the atmospheric pressure. As a result, air rushes into your lungs. When you exhale, you are simply relaxing the diaphragm muscle which compresses the lungs. As the lungs become compressed, the pressure in them rises to a point higher than the atmospheric pressure and air flows out of your lungs. The bronchioles that carry air through your lungs have elastic walls so they stretch as air flows through them. During the destructive process of emphysema, the elastic walls of bronchioles can be destroyed and the bronchioles can collapse. When a person with emphysema tries to exhale, they must work very hard by contracting chest muscles to squeeze air out of the oversized alveoli and through the collapsed bronchioles.

Emphysema is a tragic disease. To watch a loved one die of it is horrifying. Over 15,000 Americans die from emphysema annually, but this is just the tip of the iceberg. Over 500,000 Americans and almost 3 million people worldwide die from smoking-related diseases every year. In the United States, smokers have life expectancies of 10 to 15 years less than non-smokers. And the effect of cigarette smoke does not end with the smoker. Second-hand smoke is also a significant health risk. If you smoke, deciding to quit is the single most important decision you can make to benefit your health and the health of those around you.

Jack is a real person with a real cancer. Jack did not smoke and was he never exposed to chronic second-hand smoke. He will never know what caused the cancer, but his surgery was 5 years ago, and he is playing tennis and fly-fishing without any ill effects. There is every indication that the surgeon was correct about the cancer being completely removed.

Questions about this case

1. A bacterial infection in the lungs could easily occur in the process that results in emphysema. This would undoubtedly cause fluid to fill the alveolar sacks in the area of the infection. What is this called? Why does it happen? What does it do to gas exchange?

2. Patients with advanced emphysema are given pure oxygen to breathe. What is the oxygen content of normal air?

3. Does it surprise you that a smoker can have the same risk of cancer as a non-smoker if they quit for 15 years? Why or why not?

4. Smokers make up over 90% of lung cancer victims. Most die within 1 year of diagnosis. Why is early diagnosis so difficult?

5. To more forcefully exhale, people contract chest muscles. Why don't people use their diaphragm?

Questions to go deeper

1. Cystic fibrosis patients also have problems with mucus accumulation as they sleep. How is it different than "smoker's hack"?

2. What are some of the other diseases that a person has an increased risk of contracting if they smoke?

3. Is second-hand smoke really dangerous? Why?

4. What effect does smoking have on a baby whose mother smoked during pregnancy?

More Deadly than War

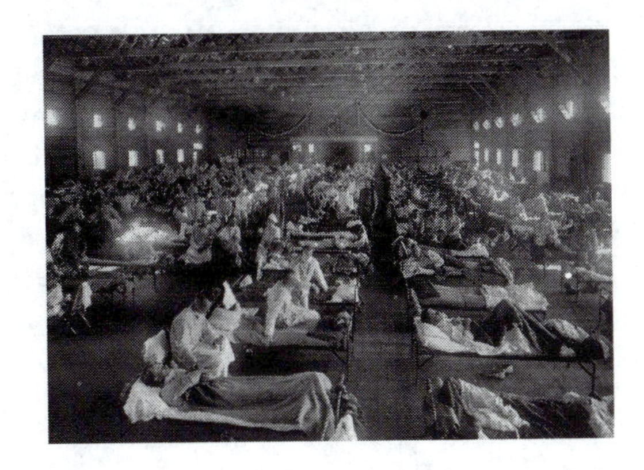

A Case Study of Influenza

The year 1918 was a momentous one. The "Great War," also known as "World War I" and the "War to End All Wars," was winding down. The United States had sent almost 4 million troops to Europe to help defeat Germany and bring peace. But, today, World War I does not capture the type of attention in the United States that other wars do. The Civil War may be more interesting, Vietnam more controversial, and World War II more dramatic. However, it was the events that occurred during World War I that forever changed mankind. Nearly 9 million soldiers died in World War I, of which about 125,000 were from the United States. The first ever wartime use of airplanes and chemical weapons was in World War I, and the war dramatically changed the politics in Europe. Yet, arguably the single most important event during the war had nothing to do with the war at all. In mid-1918, an outbreak of a new strain of **influenza**, called the **Spanish Flu**, occurred, and in a matter of a year, almost 25 million people died worldwide. The statistics from this **pandemic** are startling. Almost half of the world's population became infected, with the mortality rate in the United States 20 times that of the flu season the previous year. In some places in the world, the mortality rate was nearly 50%, and some villages were completely wiped out. Of the servicemen who died in Europe, over half of them died from influenza. No infectious disease on record can compare to the 1918 pandemic–not the Black Plague outbreaks in the Middle Ages, or HIV. The photograph above shows a typical scene from this era. At first glance, it looks like an image of any army hospital; in reality, this hospital is full of patients suffering from influenza.

It is important to understand how such a pandemic can occur, especially in light of recent problems with **bird flu.** First let's get a full understanding of the disease. Influenza, or flu, is caused by an infectious influenza **virus** that comes in three varieties named A, B, and C. While types B and C can cause disease in humans, it is type A that is responsible for the major pandemics. The virus is transmitted by air or touch through the saliva coughed out by the victim, and is highly infectious. The virus primarily infects the upper respiratory tract, and typical symptoms include cough, headache, body aches, fever, sore throat, and fatigue. To eliminate the virus, the human immune system produces specific **antibodies** to the virus **particle**. During the normal flu season, the mortality rate is rather low (about 0.1%), and mortality is usually a result of complications in victims whose immune systems are not the strongest. The **B-cells** that circulate in your blood make antibodies to foreign molecules like those on the flu particle. After the foreign molecules are eliminated, the B-cells store the ability to make more of that specific antibody by making **memory cells** that circulate through the blood. If the same protein is ever found again, the memory cell divides rapidly, making a large quantity of antibodies to eliminate it. Since the virus each year is similar to, but not the same as, the virus of the previous year, the antibodies that were made the year before are helpful in limiting the growth of the virus. This is the reason for the typical low mortality rate. Therefore, mortality is usually highest in the young and the elderly because they may not make antibodies as effectively as necessary or the symptoms of the viral infection may complicate other diseases. It is important to note that the influenza virus is not the cause of the stomach or intestinal problems sometimes called the "stomach flu" or the "24-hour flu." These problems can have a variety of causes, but should not be confused with true influenza.

If flu mortality rates are usually fairly low, you might ask, "What happened in 1918?" Answering this question is the key to understanding flu. The influenza virus has **RNA** for genetic material and then has two critical proteins in the coat that surrounds the RNA. These proteins are called **hemagglutinin** (HA) and **neuraminidase** (NA). When the virus infects human cells, it uses the machinery of the human cells to make all the parts to assemble new virus particles. To accomplish this, viral genetic information must be converted from RNA to DNA since the human cell only has the **enzymes** to use DNA as instructions to make proteins. The virus supplies an enzyme to convert the RNA to DNA. After the viral RNA is converted into DNA, the host cell machinery uses the DNA to produce viral proteins and then these proteins are assembled into new viruses. Ultimately, the cells burst, releasing the new virus particles that can go on to infect other cells. Unfortunately, the **polymerase** enzyme responsible for the conversion of viral RNA to DNA cannot "**proofread**." Proofreading is the process that some polymerases use to check their work to make sure they do not make mistakes. The human DNA polymerase that copies DNA during **replication** proofreads in order to limit the number of mistakes. Since the influenza polymerase cannot proofread, it makes more mistakes, and mistakes lead to **mutations** in the virus. A mutation may cause **genetic drift**. This means that a slightly different virus than the previous year is produced. The next flu season should not be much more serious than the previous one because the virus is similar enough to previous flu viruses. The antibodies that were made for previous influenza viruses

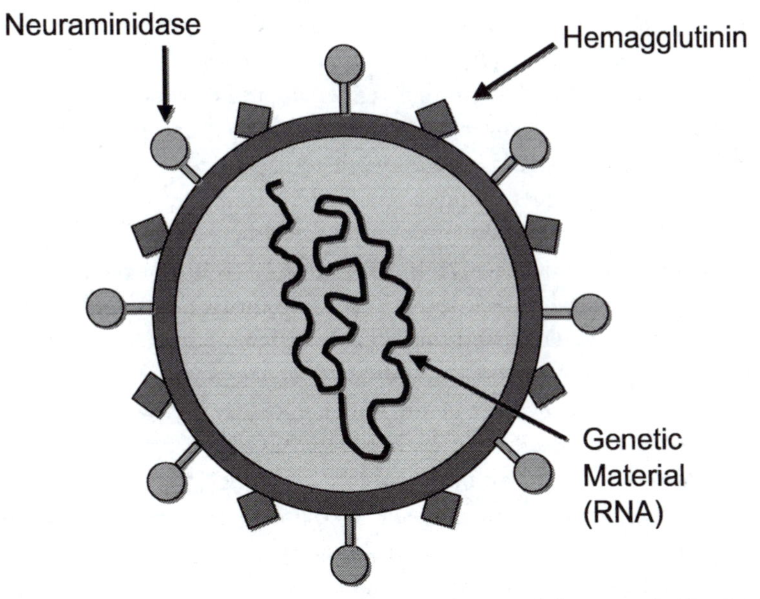

This schematic of a influenza virus shows the genetic material surrounded by the "capsid" which includes the hemaglutinin and neuramindase.

would be helpful in eliminating the new, slightly different virus. If a mutation makes a significantly different virus, then a **genetic shift** has occurred. In this situation, the human immune system would treat the virus as entirely new and previous exposure to other flu viruses would not be helpful. The genetic shift can result in the type of pandemic seen in 1918.

In light of the molecular events that change the influenza virus, research scientists have created a system to name the new virus **strains**. A particular strain might be called A/Beijing/32/92 (H3N2). This means that it is influenza type A, isolated in Beijing, the official strain number 32, and was isolated in 1992. The (H3N2) indicates the variant of the HA and NA molecules. The name is often shortened to A(H3N2). This terminology began in the 1918 outbreak so the 1918 virus is called A(H1N1). The full name would be specific for an individual virus isolated during the 1918 outbreak, something like A/Boston/6/18 (H1N1). The terminology is helpful in understanding the flu pandemics that have occurred since 1918. In 1957, a genetic shift resulted in an outbreak of flu type A(H2N2). This pandemic killed 70,000 people in the United States. In 1968, another shift caused the outbreak of the A(H2N3) virus that killed 35,000 in the United States. There has not been a major pandemic since 1968; we are long overdue for another significant genetic shift.

The influenza virus changes every year, and with every flu season, scientists are interested in what type of influenza will be present. To prepare for the flu season, many health care professionals recommend a flu vaccine. Since the exact variety of virus that will be present is unknown, vaccines are made to protect against a group of the most common influenza types. It is likely that the viruses used for the vaccine will approximate the flu virus for the coming year enough to provide some protection. In the event of a genetic shift event, we can expect little or no protection from the vaccine since the virus will have dramatically changed. Vaccines against the influenza virus are made with the same technique invented 50 years ago. The flu virus is injected into a living fertilized chicken egg, which is resealed and incubated. When the embryo becomes infected, it produces viral particles, which are harvested and turned into the vaccine. It is a laborious, time-consuming process. If a shift event were to occur, the new strain would be used to make a vaccine, but it would take so much time that most people would not benefit immediately.

How does bird flu relate to all of this? Over the past 5 years, a flu strain called (H5N1) has been discovered in poultry. This virus has been shown to be able to infect humans who come in contact with infected birds, although there is a lack of evidence for the traditional human-to-human spread of the virus from coughing up contaminated saliva. The virus has a high mortality in humans, but if contact with infected birds is the only mode of transmitting it to humans, then a human pandemic is unlikely. The concern is that as the viral RNA is replicated by non-proofreading polymerases a mutation might occur that would allow human-to-human transfer. The result could be as devastating as the 1918 pandemic. What is worse is there would probably be a simultaneous pandemic in birds, which would severely limit the number of eggs that would be available to make vaccines.

What can be done in preparation? Research is certainly important. Knowing more about the H1N1 and H5N1 strains will help us understand more about influenza **virulence** and should help in vaccine and drug development. Additionally, there is available an anti-viral drug called Tamiflu® that provides some relief during influenza infections, but is not a cure. Tamiflu® should slow the virus enough to limit mortality, so if a flu pandemic of any type occurs, it will be important for local health care agencies to have access to enough Tamiflu® to treat the seriously ill.

There is clearly concern from the current bird flu outbreaks of another influenza pandemic. Many people feel the press is trying to scare people without reason. I hope this case shows that there is reason to take influenza seriously. Some scientists fear that we are spending so much money and effort on diseases like cancer and HIV that we have forgotten the near certainty and potential severity of another serious flu pandemic.

Questions about this case:

1. Why do we call a virus a particle and not a cell?
2. What is the difference between a pandemic and an epidemic?
3. In light of how flu is spread from person to person, what precautions should you take in a normal flu season to prevent contracting the flu virus?
4. What precautions should be taken during a pandemic to avoid the flu?
5. Some people have an allergy that prevents them from being able to take a flu vaccine; which allergy is this?

Questions to go deeper:

1. Are a cold and the flu the same thing? Why or why not?
2. Where was the first case of flu in 1918 discovered?
3. Who were some of the people who had the highest mortality rates in the 1918 flu pandemic? Why?
4. Why are both the influenza virus and the human immunodeficiency virus notorious for their ability to mutate?
5. Physicians find that patients with colds or the flu often ask for an antibiotic, but antibiotics do not work for colds or the flu. Why?

References:

1. http://virus.stanford.edu/uda/
2. http://www.cdc.gov/flu/

The Professor in the Orange Shirt

A Case Study of HIV and AIDS

As a faculty member at a small university, I get the opportunity to be involved in student activities. Subsequently, I have become good friends with the folks over in the Student Activities Office, and one day last year, they called and asked if I would be willing to wear a plain orange T-shirt to work every day for the following week. Since I love to get involved in the lives of students outside the classroom, I immediately agreed. But, what was this all about? Now that I had agreed to wear the shirt, I figured they would tell me why. Our student life office was having a number of faculty, staff, and students on campus wear the orange shirts on the following Monday. The total was to represent 1/35 of the university's population. The volunteers would wear them again on Tuesday, with more added to their number so that 1/20 of the people on campus would be wearing orange shirts. They had it worked out so that on Wednesday, 1/15 of the people on campus would be wearing orange shirts, followed by 1/10 on Thursday. What did it mean? It was AIDS awareness week. This demonstration was adapted from a similar event at a sister university and represented the proportion of children in Africa who are orphans due to AIDS. In 2000, 1 in every 35 children were orphaned due to AIDS. In 2005, 1 in every 20 children were AIDS orphans, and if nothing is done to stop the tragedy by 2010, 1 in 15 children will be orphaned. By 2015, 1 in every 10 will be orphaned! That's a projection of nearly 30 million orphans by 2015! It was quite a visual statement when you walked around campus and saw all the orange shirts. Initially, students had no idea what they meant. I had dozens of students ask, and when I explained, they were astonished.

AIDS is short for **acquired immune deficiency syndrome**. AIDS was first discovered when people began to be admitted to hospitals with infections that were considered rare and **opportunistic**. This means that people normally are not infected by these organisms or the infection results in little pathology unless the person has a compromised immune system. For example, ***Pneumocystis carinii*** is a **protozoan**

parasite that only infects people with poor immune systems and has catastrophic effects. Infection with this parasite can result in severe, and often fatal, pneumonia if the body cannot kill the parasite. People at the **Centers for Disease Control** (CDC) became interested when there was a significant increase in the occurance of opportunistic infections. A **syndrome** is simply a collection of signs and symptoms. All the signs and symptoms that are associated with AIDS seemed to be the result of a significantly damaged immune system. Initially, scientists did not know the cause. The first cases of AIDS were diagnosed primarily in homosexual men. But, it was quickly established that IV drug users and people who received blood transfusions were being diagnosed with the same problems. In 1984, a group of researchers at the Pasteur Institute in France discovered a virus they called LAV, which they claimed caused AIDS. A year later, researchers at the U.S. National Cancer Institute discovered a virus they called HTLV-III, which they claimed caused AIDS. It turns out that both research groups were studying the same virus, which was later called **human immunodeficiency virus,** or **HIV**.

We are infected by viruses all the time, so why does this one cause such significant harm? HIV is a **retrovirus,** meaning that its genetic material is **RNA** (not **DNA** like most organisms). It, therefore, must convert its RNA into DNA upon infecting a cell by using an enzyme called **reverse transcriptase**, which is encoded by the virus **genome** and is carried in the virus particle. When HIV infects cells, it can progress through two different growth patterns. Once the RNA is used to make DNA, the new DNA can be incorporated into the human genome (using a virally encoded protein called **integrase**), where it can remain latent. In this situation, the cell functions normally. But, if the virus becomes active, it will commandeer the cellular machinery and use its own genetic material to make the proteins required to assemble many new virus particles. The virus particles are then released, killing the cell in the process. HIV infects primarily cells of the immune system, such as **glial cells**, **macrophages**, and T-cells. AIDS is specifically the result of infection of specific T-cells called the **$CD4^+$ T-cells**. $CD4^+$ T-cells are a subset of T-cells that have the CD4 protein on their surface, which indicates that these cells are what we call **helper T-cells**. HIV actually utilizes the CD4 molecule to enter the $CD4^+$ T-cells. As the HIV infection progresses, the number of $CD4^+$ cells a person has can drop significantly. To understand the significance of the loss of $CD4^+$ cells, a brief overview of the immune system is in order.

The immune system is divided into two parts: the **cell-mediated immune system** and the **humeral immune system**. The humeral immune system consists of the cells required to make antibodies. **B-cells** produce specific **antibodies** to neutralize foreign molecules, which are generally called **antigens**. An antigen that has an antibody attached is a target for elimination by cells of the cell-mediated immune system. Once the foreign object is removed, some memory cells are formed. These cells retain the memory of how to make a specific antibody, so if a specific antigen is found again, large amounts of antibody can be produced very quickly. On the other hand, the cell-mediated immune system consists of a variety of different cells that can identify and eliminate foreign objects. Some of these cells have the ability to produce toxic molecules to kill invading microorganisms, while others can simply engulf invading particles. The latter of these two are called **macrophages,** due to their enormous appetite. Both groups of these cells make up an army that becomes active when they receive the appropriate signals. These signals come in the form of **cytokines** released from the $CD4^+$ T-cells. As the HIV infection specifically kills the $CD4^+$ T-cells, the rest of the army becomes unable to fight off infections because the molecules that activate them are absent. The very cells that are present to organize the attack against HIV are specifically killed by HIV.

Over the years, the misinformation regarding how a person can become infected has led to horrible injustice. In the 1980s, fear overcame reason. People feared hugging, shaking hands with, and even living next door to people who were **HIV+**. HIV can only be spread by contact with a bodily fluid from an infected person, and even if contact occurs, it does not guarantee infection. With rare exceptions, HIV has been transmitted almost exclusively by unprotected sex, sharing of needles by drug users, an infected mother to her child during pregnancy and birth, and blood transfusions. The current testing of the donated blood supply has eliminated transfusions as a risk for HIV transmission. It is important to remember that being HIV+ and having AIDS are very different. A person can live many years while being infected by the virus (HIV+) but show no symptoms of the infection. Only when the virus kills a significant proportion of the $CD4^+$ T-cells do the symptoms of AIDS show up. Therefore, it is possible for a person to be infected and infectious without ever knowing they are HIV+. We would call them **asymptomatic**, or **carriers**, and they have played a major role in the spread of HIV.

How are HIV infections treated? The goal in treating any infection is to damage the infectious agent without hurting the patient too badly. Scientists often look to block molecular processes that are essential for the infectious agent but not for the host. HIV requires the enzyme reverse transcriptase to infect cells, but the human cells never have a need for the enzyme. Initial attempts in drug intervention for those who were HIV+ were to treat them with drugs that inhibited the action of reverse transcriptase. The drug **AZT** was the first one developed for this purpose, and many scientists believed that treatment with AZT alone could stop the infection. They were wrong. Over many years, different drugs have been developed, many of which inhibit reverse transcriptase. Many regiments have been developed for patients with differing circumstances. These drugs can be quite expensive, and, therefore, are not available to patients without the means to pay for them. Even with the drugs, we have not found a cure for HIV infections.

With no cure for HIV, scientists have been working at preventing infection. There is significant interest in a search for an HIV vaccine. The goal with a vaccine would be to treat HIV-people with some antigen from HIV so that the B-cells would make an antibody that would protect against HIV infection. After the vaccine, the memory cells would develop, and if an infectious HIV were ever encountered it could be effectively eliminated. A vaccine would be considered useful even if it protected only some individuals or delayed the progression of HIV infections into AIDS. Many attempts at creating vaccines have been made with little success. Scientists continue to work feverishly, yet currently there is no HIV vaccine. But, prevention means more than finding a vaccine. Since most infections are currently sexually transmitted, it follows that we could almost rid the world of AIDS in one generation if people universally practiced safe sex. Ultimately, the safest sex is abstinence. Your chance of becoming HIV+ by having sex using a condom is small, but your chance by abstaining from sex is almost zero. Worldwide education regarding how HIV is transmitted and how to avoid HIV infections is a key piece to slowing the spread of this disease.

To conclude, I thought some statistics would be in order. These are from the world relief organization, World Vision. There are more than 40 million people worldwide who are HIV+, and there are almost 15 thousand new infections daily. The average life span in Botswana in 2010 will be 27, but would be 70, if it were not for AIDS. Sadly enough, we spend more money on Viagra every year here in the United States than has been donated to the Global AIDS Fund since 2001.

Questions about this case

1. What is the difference between having AIDS and having HIV (being HIV+)?
2. Why do we call a virus a particle and not a cell?
3. There are actually two types of HIV: HIV1 and HIV2. How do they differ?
4. What are some of the other opportunistic infections that are part of the AIDS diagnosis?
5. The test that has been developed for detecting HIV infections in humans does not test for the actual virus since there may only be a small amount of it in blood. What does it test for if not for the actual virus?
6. Four "nearly exclusive" means of transmission of HIV are mentioned in the text. Name one rare, confirmed transmission not included in the text.

Questions to go deeper

1. We mentioned $CD4^+$ cells, but not $CD8^+$ cells. What is the $CD8^+$, and what does it do?
2. What are the glial cells mentioned in the third paragraph?
3. What is Kaposi's sarcoma?

References

http://www.avert.org/his81_86.htm
Acting on AIDS Statistic Sheet, 2004. www.worldvision.org

Mom vs. the Botanist

A Case Study of Plant and Fruit Anatomy

"Okay, class. In today's Botany lab, we are going to be looking at fruit."

"Where are the fruit, Dr. K.? All I see are a bunch of vegetables like these tomatoes, cucumbers, beans, and an avocado."

"They are all fruit, so let's get going!"

"These aren't fruit! These are vegetables! That's what my mom told me!"

Well, without calling her mom a liar, I proceeded to explain the difference between what her mom and the produce guy call things, and what a botanist calls things. In **botany**, the science of plants, there is no such thing as a vegetable. They are just different plant parts.

If we look at plants from the ground up, we would start with the **roots**. These are the parts of the plant that anchor the plant, absorb water and minerals, and, frequently, store food for the plant. Yams and sweet potatoes are roots that store material that, if we didn't rip them away from the plant and eat them, would be used for the next season's growth. That is what makes them so yummy to eat at Thanksgiving.

Everything above the ground in a plant is part of the **shoot** system, and we eat lots of shoot parts of plants. Before we look at those, however, let's look at some good things to eat that are part root and part shoot. Carrots, beets, and radishes, for example, are a little root and a little shoot. The larger part that we actually eat is primarily root, while the top of the edible root is really made of shoot material. If you have ever seen these growing, you'll know that part of the carrot, beet, or radish actually sticks out of the ground and has the leaves attached to it. When you pull it out of the ground, you can see the stringy lateral roots running off the main taproot, which is the part we actually eat.

Besides these "half and half" parts, most things we eat are parts of the **stem**, **leaves**, **flowers**, **fruits**, or **seeds** of the plant. The stem is the main part of the plant that supports the rest of the shoot system. In trees, it becomes the trunk and branches that we climb on. "Vegetables," such as asparagus, celery, and, believe it or not, white potatoes are all actually stems. That might seem okay as far as the first two go, since they are obviously above the ground, but a potato? Everyone knows those grow underground. How can they be stems? Well, there are lots of different types of stems and some of them run along the ground. If they are above the ground, they are called **stolons**. If they are below the ground, they are called **tubers**, which is what a potato is.

To further confuse things, if you still think all stems should be above the ground, what about leaves? Certainly they all must be above the ground! Well, not always. Think about onions. Those are actually leaves and, once again, if you have ever seen them growing, you know that the white things that we eat are actually attached to green leaves that stick up above the ground. In this case, however, the white part is where the excess food is stored. Other things that we eat, such as cabbage, spinach, and lettuce are more traditional leaves. They are above ground, they are green, and they actually look like leaves. In fact, if you pull them apart leaf by leaf, you can see that they are each actually attached to a short little stem just like a regular leaf is attached to the branch of a plant. Where the leaf attaches to the stem is called a **node,** while the space in between is called an **internode**. In the case of cabbage, spinach, and lettuce, the internode is really short so the leaves all bunch together.

Pretty much everything else we call vegetables are some part of the reproductive system of a plant. In the group of plants called **flowering plants,** this all starts with the **flower**. While we associate these with the colorful parts of the plant, the parts that we give to our sweeties, some of them can also be eaten. For example, cauliflower and broccoli are actually the flowers of these plants. If you let these sit too long in the refrigerator, you can actually see them bloom, so they do look a little more like real flowers.

A flower may contain only male parts, only female parts, or both. The male parts are known as the **stamen** and consist of the slender **filament**, which is topped by the enlarged **anther**. This is what produces the **pollen**, which contains the **sperm** of the flower. For the pollen to get to the female part of the flower, or the female flower itself, a process known as **pollination** has to occur. The colorful parts of the plant are designed to help this process occur. Usually these **petals** are brightly colored to attract birds and insects that come to get food from the flower and, in doing so, pick up pollen that they then carry to the female part. If the pollen gets to the female part of the flower, called the **pistil**, **fertilization** will occur. This happens when the pollen falls onto the sticky **stigma**, germinates, and grows down the **style**, and enters the bottom part called the **ovary**. This is where the egg is and it is well protected from drying out and it is one of the reasons these types of plants are so well adapted to living on land.

Once the egg is fertilized, it develops into a **seed**, which, if we plant it correctly and water it, will become a new plant. Unless we eat them, of course! If you have ever eaten a kernel of corn, a pea, a peanut, or an almond, you have really eaten the seed of a plant. However, you do not have to worry about these seeds germinating and growing in your stomach as your older brother may have told you when you ate those watermelon seeds! The action of chewing, your stomach acid, and the enzymes in your digestive system all serve to destroy the seed and prevent it from growing.

To protect the developing seed, parts of the flower begin to enlarge and become the **fruit**. Most of the time it is the ovary that enlarges but, in some cases, it is the base of the flower, called the **receptacle**, that starts to grow. This is why, by definition, a fruit is anything containing seeds. That is the reason so many of the things we call vegetables are really fruit. But it isn't that simple. Botanists divide fruits into several categories based not upon what they are like when we think they are ready to eat, but what they are like when they are ready to release their seeds, the final role of the fruit. Therefore, the two main categories are **fleshy** and **dry** fruits. These categories refer to the condition of the fruit when they release the seeds. If the fruit is still soft and fleshy and turning into mush when the seeds are released, it would be a fleshy fruit, but if it has dried and begins to split open to release the seeds, it would be a dry fruit.

Fleshy fruits include things that both a mom and a botanist would agree on, like apples, oranges, peaches, grapes, and cherries, and lots of things they would not, like squash, olives, tomatoes, cucumbers and avocados. However, if you think about it, all of these really do have seeds inside them, even if it is just one big one, and that is what makes them fruit. Within fleshy fruit there are lots of subcategories like **pepos**, **pomes**, and **berries**. Once again, however, things can get confusing, as several things we call berries are not botanical berries. Strawberries, blackberries, and raspberries are not berries at all, but oranges and tomatoes really are!

Dry fruits are broken down by whether or not they split when they dry out and how in many places they split if they do. All kinds of beans, pea pods, peppers, chilies, and nuts are examples of dry fruit. As you would probably have guessed, however, some of these things are not what you think they are. This is particularly true of nuts. Examples of true nuts include acorns, filberts, and hickory nuts. Most things we call nuts are not nuts, according to botanists. Peanuts, coconuts, walnuts, pecans, cashews, and pistachios are all examples of plant parts we eat that are not true nuts.

So, if all of these are fruits because they contain seeds, what do we do about those seedless varieties of fruits we buy? Does the fact that they lack seeds make them something else? Well, no, because they are formed from flowers and, in many cases, would contain seeds if humans did not mess with them. In fact, whether they do it naturally or because of human manipulation, we do have a number of seedless varieties of fruit...but that's a whole other story.

Questions about this case:

1. The case says a strawberry is not actually a berry and a tomato is. Okay, then, define what makes a berry a berry!
2. Again, if a peanut and a coconut are not nuts, then what is a nut?
3. White potatoes and onions both store excess food for the plant. What specific type of food does each store?

Questions to go deeper:

1. When you go to your local nursery you can find fruit trees that are "self-fruiting" and some that are not. How are these trees different?
2. Besides storing food, some plant parts store water. What are some examples of plant parts that do this?

Today's Forecast: Chance of Thunder Showers in the Produce Section

A Case Study of Osmosis, Transpiration, and Plants

The produce section of the local grocery store is not the place you expect to hear thunder, but that was what I was hearing. The closer I got to the heads of lettuce, celery, and cucumbers, the louder the rumbling was. My curiosity got the best of me, and I paid for it! As I stuck my head into the display case to see where the sound was coming from, I was rewarded with a shower, as the mister running along the top of the display opened up, sending a fine spray of water over me and the produce. Feeling like a fool, I realized that the sound of thunder was the store's way of warning customers that the produce sprinklers were about to come on. They apparently did not take into account the effect it would have on a too-curious biology professor!

While the thunder sounds might be a little unusual, I am sure you have noticed the spray heads found in most produce sections. What you may not have realized is why they are there. The reason is tied into why your celery goes limp after being in the refrigerator, why your refrigerator has a "crisper" drawer in it, why plants wilt, and why 350-ft. redwoods can get water to their leaves without electricity, pipes, and pumps. It is all based upon what is happening at the cell level, regarding cell structure, and how water moves across the cell membrane.

Remember that water is one of the most important molecules in a cell and that it makes up about 90% of the cell's interior. If water is that important, the cell must have a way to get it to move back and forth across the cell membrane and cell wall. This process is called **osmosis** and it is a variation of **diffusion**. Diffusion is defined as the movement of molecules from areas of high concentration to areas of low concentration, and it occurs all around us. If you have ever walked into a room where something smelly is sitting in a corner and noticed that it soon could be smelled throughout the room or seen your sugar dissolve in your coffee without being stirred, you have experienced diffusion. Unless they are at **absolute zero**, all molecules vibrate a little. When they vibrate, they bump into the molecules around them. When they bump into each other, they move apart and bump into something else. Obviously, if you have a lot of molecules near each other (high concentration) they will bump into each other more often, and that will cause them to spread out. Eventually, they will be pretty evenly spread out over the system.

Now, if you add a **semipermeable membrane** that allows only water to move across it, you have osmosis. This is basically just diffusion of water across a semipermeable membrane. So what does "semipermeable" mean? Imagine you are in a room with your professor on one side of a giant volleyball net and you are on the other side. As long as you throw beach balls, basketballs, volleyballs, and other large balls at him, he is safe because the net will not allow those to go through it. However, if you start throwing golf balls, marbles, ball bearings, etc., he is in trouble. Now, I am not suggesting that you try this experiment, but you get the idea. If an item is small enough, it can get through a semipermeable membrane. If it is too large, it cannot. Cell membranes are like the volleyball net, as they will keep some molecules out and allow others, like water, through.

But what determines if the water goes into the cell or outside of the cell? That is where the water concentration comes in. Water is the **solvent** that has solid material, the **solute**, dissolved in it to create a **solution**. Most cells have a salt concentration of 0.9% NaCl. That means, ignoring all the other stuff in the cell, the water concentration is 99.1% H_2O. If we have a solution that is 10% NaCl outside the cell, then there is 10% salt and 90% water on the other side of the cell membrane. In cases dealing with osmosis, we assume that the only thing capable of moving across the membrane is the water. In this case, the water is going to move from where it is highest (99.1%) to where it is lowest (90%) and, therefore, water will leave the cell. We say that this cell is in a **hypertonic** solution because the NaCl concentration is greater (hyper-) outside the cell than inside the cell. If the situations were reversed, then the cell would be in a **hypotonic** solution, with the NaCl concentration being less (hypo-) outside the cell than inside the cell. In this case, the water would move into the cell because there is more water outside the cell (99.1%) than there is inside the cell (90%). The final possibility is where the concentrations are equal inside and outside the cell. This is called **isotonic** (iso = equal), as there is the same concentration of H_2O and NaCl outside the cell as inside the cell. Now the water molecules will move randomly in and out, but there will be no net increase of water inside or outside the cell.

So what does all this have to do with limp celery and wilting plants? If you remember, plant cells have a **cell wall** around the outside of the cell membrane. These cell walls provide the support that keeps the plant and its parts upright. However, this cell wall is not a solid rigid box around the cell. It is a living, flexible unit that is further supported by the water pressure inside the cell. If water is leaving the cell because it is in a hypertonic solution or condition, then there is no **turgor pressure** pushing against the cell wall and the support is not there for the plant. This causes the celery to go limp or the plant to wilt. As soon as you add water to the plant's environment, by placing the celery stalk in water, by spraying the produce with a mister, or by watering the plant, it is in a hypotonic condition. Water thus moves into the cell, fills up the cell interior, and exerts tremendous turgor pressure on the cell walls. This makes them much stronger and able to support the plant. The celery is nice and firm, the produce is crisp, and the plant does not wilt.

While this explains several of the questions we asked at the start, this is only a very small part of how water is able to move from the roots, up through the stem, out the branches and to the leaves. While scientists are not totally sure of how this happens, they have a pretty good theory. While an animal has blood vessels and a pump—the heart—plants lack these features. While a building has pipes, pumps, and water pressure from the city, plants lack these features. So how can they get water up to the top of a tree that is hundreds of feet tall? Instead of vessels and pipes, trees have **vascular tissue** called **xylem** and **phloem**. The xylem carries water and nutrients up from the roots to the leaves, while the phloem carries the sugars from photosynthesis down to the roots. Instead of a pump pushing material up the pipes or through the body, however, plants rely on a complex system of processes. Ultimately, it is the process of **transpiration** that provides the power or pull that drives the process. Transpiration is the evaporation of water from any type of vegetation. This occurs primarily through the leaves of the plant, which have tiny openings called **stomata** that serve as areas where gases such as CO_2, O_2, and water vapor can enter and exit. Since water is leaving the leaves, it creates a constant need for more water to replace the water that is leaving. Since water molecules stick to each other and to the sides of the xylem, this means that as water leaves the plant, it pulls the next water molecule up which pulls the next water molecule up, and so on and so on. The ability of water molecules to do this is due to the **semipolar** nature of water molecules. Due to the **covalent bonds** associated with the molecule, electrons are shared between the oxygen atom and the two hydrogen atoms. However, the larger oxygen nucleus pulls the shared electrons more than the smaller hydrogen molecules, so the oxygen end is slightly negative and the hydrogen side is slightly positive. If you have a whole bunch of these molecules together, the slightly negative oxygen side of one water molecule will form a weak **hydrogen bond** with an adjacent slightly positive hydrogen side of another water molecule. The slightly negative oxygen of this water molecule will then form another hydrogen bond with the slightly positive hydrogen of another water molecule, and so on and so on. This is basically what is happening when you sip water through a straw, only your sucking on the end of the straw is what provides the "pull" instead of transpiration. When water has to move across cell membranes and cell walls in the system, diffusion and osmosis drive the movement. Again, if water is leaving one cell, it creates a situation where the water has to be replaced by water from the next cell. Thus, the water slowly moves from the soil into the roots, through the roots and up to the stems, then out the branches and out into the leaves through an amazingly complex system of structures and processes.

This idea of osmosis in cells is not restricted to plants, however. In fact, this is a crucial issue for people suffering with **hypertension,** or high blood pressure. In many cases, their high blood pressure is partially due to the fact that their diet is too high in salt. As the body absorbs this excess salt, it changes the osmotic concentration of the body, creating a hypertonic environment. In order to compensate for this, the body keeps or retains more water in an attempt to bring the body back to an isotonic situation. Unfortunately, that means there is excess water in the body, which increases the blood volume, causing high blood pressure. If this condition continues for too long, significant health problems and death may result. Therefore, it is important that a person with high blood pressure lower his or her consumption of salty foods.

Questions about this case:

1. How does the organelle called the central vacuole in plant cells play a role in maintaining turgor pressure?

2. Suppose you place two celery stalks with cut ends in a solution of red food coloring and remove all the leaves from only one of the stalks. If after 30 minutes you remove the celery and cut the stalks to see how far up each stalk the red dye has traveled in the plants' vascular systems, in which stalk should the red dye have moved the farthest? Why?

3. If water is drawn up a tree's vascular system by virtue of its ability to form hydrogen bonds, it would make sense that the height of trees would be limited by the strength of these bonds. So, how high is the tallest tree? What are scientists saying about transpiration and the limit of the height of trees?

Questions to go deeper:

1. One would think that desert plants would want to limit transpiration. How do desert plants conserve water and limit loss?

2. As mentioned in the case, water molecules stick together. It is really an issue of hydrogen bonding. What are a couple of other characteristics of water that are a result this "stickiness"?

3. Cell membranes are semipermeable. The case suggests that water can pass through such membranes easily. What are some types of molecules that cannot diffuse across cell membranes? How might these molecules cross the cell membrane if they cannot diffuse?

If You Think Telling Your Kids About Sex Will Be Hard...

A Case Study of Birth Control

"Daddy! I can't talk to you about this! That's just too weird!"

My daughter was about to graduate from college, working about 20 hours a week, and planning her wedding. I knew she was pretty busy and I was afraid that the last thing on her mind would be planning on the form of birth control they were going to be using. I knew from past experience with students and friends that people frequently waited until it was too late because they did not realize how long it took to get on some forms of birth control or to have them become effective. I didn't want that to happen to her but, I have to admit, it was not a discussion I was looking forward to.

Fortunately, being the daughter of a biologist, she had a pretty good background in human anatomy and physiology, which is really all that is required to understand the various forms of birth control. Let's start by considering the male of the species as, currently, there are only two forms of birth control that the male can use. Males produce their **sperm** in the tiny tubes called the **seminiferous tubules,** which lie inside the **testicles**. These organs are inside the **scrotum**, which allows them to be outside the main part of the body. That is necessary because normal body temperature is too high for the sperm and will eventually kill them. This is one reason why men who are overweight or who wear tight-fitting underwear or clothing frequently have trouble conceiving, as their testicles are too close to the body and get too warm. Another problem along these lines can be overuse of hot tubs, especially when a couple is trying to conceive. In fact, while it is not practiced too much nowadays, in many countries, women would knit little pieces of wool to cover the scrotum as a form of birth control!

Besides making the sperm themselves, the testicles are also the sites of **testosterone** production. The production of the **gametes** and the testosterone are under the control of the **follicle-stimulating hormone (FSH) and luteinizing hormone**

(LH). These are produced by the **pituitary gland**, a small gland located at the base of the brain just behind the nose. In turn, this gland is under the control of the **hypothalamus**, a part of the brain. In one of the best examples of how the body uses **negative feedback control**, the hypothalamus produces a hormone called gonadotropin-releasing hormone (**GnRH**) whenever the testosterone level gets too low. This causes the pituitary gland to release FSH and LH, which cause the testicles to produce sperm and testosterone. As the testosterone level gets higher, it shuts down the hypothalamus until the testosterone level gets too low again.

However, as long as the sperm are being made, they will be released from the testicles into the **epididymis**, which lies on top of each testicle. This is the site where the sperm mature, where excess or deformed sperm are absorbed, and, to an extent, where they are stored. From here they travel to the **vas deferens**, where they are stored until they are released during ejaculation or reabsorbed to make room for fresh sperm.

The final portion of the male reproductive system is the **urethra**, which functions in both the reproductive system and the urinary system. Most of the time, the urethra, which runs the length of the **penis**, is used for urinating. However, during intercourse, a tiny sphincter muscle opens at the end of the vas deferens to allow the sperm, now combined with **seminal fluid** (which comes from several accessory glands), to flow out as **semen**. It is at this point that the most common form of male birth control, the **condom**, is effective. The condom is simply a piece of latex or other material shaped to fit snugly over the penis so that it catches the semen as it is released and prevents it from entering the **vagina**. Since the condom also prevents physical contact between the male and female reproductive parts, this is the only form of male birth control that can prevent the spread of **STDs (sexually transmitted diseases)**.

A look at the female reproductive system shows us that, while the parts may have different names, the set-up is pretty similar to the male's, and some of the same hormones are involved. However, there are a great many more forms of birth control currently available for females. Just like males have two testicles, females have two **ovaries** that are the site of **egg** formation. These organs rest in two little depressions in the female's pelvis, where they are well protected. The ovaries, like the testicles, are also the site of hormone production, as some of the **estrogen** and **progesterone** are produced here. Since the female hormone physiology is more complex than the male's, we will look at the physiology of this system at the end of this section.

Each month between **puberty** and **menopause,** one of the ovaries produces and releases an egg. This process is known as **ovulation**. As the egg is released, it leaves the ovary and crosses a small gap before entering the tube called the **oviduct**, **fallopian tube**, or **uterine horn**, depending on whom you talk to. These tubes are lined with **ciliated epithelial cells** that produce a beating action that creates a current that moves the egg into the oviduct and then along into the next structure. This process takes about a week on average and, since the egg is only viable for a day or two, if fertilization is going to take place, it will have to occur in the oviduct, as the egg will be dead by the time it reaches the **uterus**. Since the oviducts are relatively thin-walled organs, they are not designed for a fertilized egg to implant in their walls. If this occurs, we call it a **tubal** or **ectopic pregnancy**, which is very dangerous for the mother and the developing **embryo**.

Whether it is an unfertilized or a fertilized egg, it moves out of the oviduct and into the **uterus**. This is the largest organ in the reproductive system, as it is designed to be the location for **implantation** of a fertilized egg and growth and development of the baby. Thus, the organ has a very thick wall composed of **smooth muscle** that is capable of expanding to hold the baby, while still providing the strength to expel the baby during labor and delivery. A highly vascular layer called the **endometrium** that is designed to support the developing embryo and, ultimately, to provide the maternal portion of the **placenta** that will support the developing **fetus,** also lines it. As will be seen below, if implantation does not occur in a given month, this layer must be shed so that a new layer can be created for the next month. The shedding of the endometrium is called **menses**, more commonly known as a period.

Several types of birth control methods have their action in the uterus. If the hormone level provided by some forms of the female **pill** does not prevent ovulation, they will prevent a fertilized egg from implanting. Implantation can also be prevented by the placement of an **IUD (intrauterine device)** in the wall of the uterus. While scientists are not exactly sure how this device works, it apparently prevents the fertilized egg from implanting by irritating the endometrium or by releasing certain chemicals. For years these devices were very popular as, once a doctor placed one into a female, they remained effective for quite a while, requiring no real attention from the female. However, in the 1970s, problems began to arise with most types of IUDs, and they were taken off the market in most countries. Recently, advances in IUDs have led some doctors to recommend them for their patients.

Females need to be aware, however, that anything that works in preventing implantation may pose an ethical dilemma for them. While there is much controversy over when life begins, people who believe that it occurs at the point of conception need to be aware that any form of birth control that works by preventing implantation results in that life ending.

If the egg is unfertilized, then it will pass out of the opening of the uterus, called the **cervix**, and into the vagina. This muscular, convoluted organ is not as thick as the uterus, but it has the ability to stretch during intercourse, labor, and delivery and to contract slightly after intercourse to assist the sperm in moving up and into the uterus. It is thus the site of sperm release during intercourse and, therefore, there are a number of birth control methods that have their action here. Some of these are common to the United States, while others seem to be more popular in other parts of the world, so you may not be as familiar with them.

A **cervical cap** is a plastic device that is fitted by a doctor to a female's cervix. It fits quite snugly and serves as a physical blockage to sperm trying to get up to the egg. This action is not to be confused with how the **sponge** or **diaphragm** works. These devices are intended to simply hold **spermicide** in the path of the sperm so that, as they swim by on their way to the egg, the chemical kills them. The sponge is impregnated with the spermicide and can be left in place for several days. The diaphragm, which also has to be properly fitted to the woman by a doctor, is simply a shallow latex cup that holds the spermicide in place. Because **spermicidal jellies, creams**, or **foams** can easily move away from the cervical opening, they should not be relied upon without the use of a sponge or diaphragm. Finally, a **female condom** has also been developed that is a piece of latex that is inserted into the vagina to serve as a receptacle for the male's penis. While not as popular as the male version, it too serves to prevent both conception and transmission of STDs.

The most popular form of female contraceptive involves manipulating the female's hormones in a way that prevents ovulation. Just as in the male, the hypothalamus releases GnRH, which causes the pituitary gland to release FSH and LH. In the female, these travel through the blood to the ovaries, where they, respectively, cause the production of eggs and estrogen and progesterone. As in the male, when the levels of these hormones get too high in the blood, the hypothalamus detects this and stops producing GnRH, which stops the pituitary gland from producing FSH and LH until the level gets too low, when the process starts over again.

If a woman does not get pregnant, then her hormone levels drop, which causes the endometrium to be shed during menstruation. However, if the woman gets pregnant, the implanted embryo starts to produce human choriogonadotropin (**HcG**), which keeps the levels of progesterone and estrogen high. As long as these levels stay high, there is no reason for the release of more FSH and LH, so no new eggs are produced.

Birth control mechanisms that manipulate this system contain progesterone, estrogen, or some combination of the two. This causes the hormone levels in the blood to become artificially high, which causes the hypothalamus to think the body is pregnant so, again, no new eggs are produced. Obviously, the most common way for these hormones to be given is through the oral pill form. However, there are also patches, implants, and other ways in which it can be delivered. If you remember back to our discussion about the male hormone regulation, you can see how similar it is to what occurs in women. Therefore, scientists are working on the sequence of events in the male to see if they can manipulate it, as we do in females, to create a **male pill** that will prevent the production of sperm.

So, while the complexity of these systems allows numerous ways to use either the anatomy or the physiology to prevent pregnancy, remember that most of the methods require advanced thought and action. Doctor visits must be arranged, follow-up visits must occur, hormone levels must be adjusted and, at times, several months may be required for effectiveness. Waiting too long could cause your dad to become a grandfather a lot quicker than he planned.

Questions about this case:

1. Some people use what is called the calendar version of the "rhythm method" as birth control. What is this and how does it work?
2. What advantage does the birth control patch have over the pill?
3. What is the withdrawal method of birth control? How effective is it?

Questions to go deeper:

1. What are the male and female forms of permanent, surgical birth control? How effective are they and can they be reversed?
2. Men often are encouraged to have a prostate exam as they get older. What is the prostate and what is the purpose of this examination?
3. A more refined form of the rhythm method is called the basal temperature method. How does this technique work, and why is it more effective than the traditional calendar method?

It's Tough to Teach Old Docs New Tricks

A Case Study of Ulcers and the Digestive System

Medical schools have taught using case studies for years, and I can distinctly remember sitting in a section of a pathology course where the professor was describing a patient and his symptoms. The patient had complained of persistent stomach pain but, the way the professor described it, I certainly wasn't thinking it was anything serious. However, as the professor read along, he came to the part that is hard for me to forget. All of a sudden, he said, the patient began to vomit large amounts of bright red blood and then died! Boom. Just like that. That particular lecture was about ulcers, which was the cause of the demise of the patient the professor read about, and I have never forgotten about how something we usually think of as being no big deal, could be.

With records dating back to the early Egyptians and published reports starting in the 1500s, it is obvious that **peptic ulcers** have been plaguing humans for centuries. However, it has only been in the last few years that modern medicine has come to the realization that we've had it wrong all these years. Since the early 1900s, doctors have thought that stress and diet were responsible for the typical ulcers that occur in the digestive system. The term "ulcer" is used to describe any open sore on the **epithelial** layer covering the outside or inside of your body or its organs. The type we are looking at occurs specifically on the lining of various parts of the **digestive system**.

One type of ulcer is called an **oral ulcer** and occurs on the epithelial lining of the first part of the digestive system, the **oral cavity**. This is composed of the mouth with its tongue, teeth, and salivary glands. The **tongue** is used for speech and the sense of taste, mixing the food and saliva into a **bolus,** which can then be swallowed. The taste buds are located on the surface of the tongue and are specialized neuron endings that are sensitive to different chemicals in the food and drink we consume. The **teeth** of an adult include four distinct types—incisors, canines, molars, and premolars. Each is designed for a specific function, such as tearing, biting, crushing, or grinding.

While oral ulcers are a true type of ulcer, they can be due to anything from biting your cheek to a viral infection like **herpes simplex**, or a dozen other possible causes. While they are painful, however, most are temporary and more bothersome than dangerous. They are not connected or related to the more significant **peptic ulcers** occurring in the stomach or small intestines.

When the bolus is swallowed, it moves down the throat to the **esophagus**, which functions simply as a tube connecting the mouth and the **stomach**. The food is moved downward through **peristalsis**, which is a common type of movement in the body. It is due to the wave-like contraction of **smooth muscle** located in the walls of an organ, which pushes the bolus ahead of it and down into the stomach. At the base of the esophagus is a **sphincter** muscle that opens and closes to allow the bolus to go into the stomach but prevents food from, normally, moving back up the esophagus.

The stomach is also a muscular organ that, like most of the digestive glands, is composed of several distinct and important layers. Remember that an **organ** is a structure composed of several different types of **tissues**, each composed of cells. The first tissue layer in the stomach is the **mucosa**, composed primarily of **epithelial** tissue. This type of tissue lines all your organs and, in this case, serves to protect the stomach from the strong acidity in the **gastric juices** as well as to secrete these gastric juices. Beneath the mucosa is the **submucosa**, which is composed primarily of **connective** tissue and **nervous** tissue. Here are found the blood vessels, nerves, and support tissue. The third layer is composed of three layers of **smooth muscle** tissue and is called the **muscularis** layer. This layer is what causes the churning of the stomach that helps mix the food and gastric juices, frequently causing your stomach to growl. Finally, another layer of epithelial tissue surrounds the entire organ. This **serosa** layer protects the organ and separates it from the body cavity and organs outside the digestive system.

When it comes to **stomach ulcers**, for decades the belief was that diet and stress created a situation in which the mucosa layer secreted an abnormal amount of **gastric juices** into the stomach. Since the principle component of these juices is hydrochloric acid that can lower the pH to 1.5, it seemed reasonable that this extreme acidity caused the destruction of the protective mucous layer. We know certain foods can cause irritation of the stomach lining and that when the body is under stress, the digestive system shuts down so that the body can concentrate on the actions needed to handle the stressful situation. During this time, the gastric juices "build up" and, when the situation returns to normal, the excess juices are dumped into the stomach, causing a sudden increase in acidity that can overwhelm the mucous layer.

As the mucous layer is destroyed in a particular area, an ulcer was thought to develop and, as the acid ate through the mucosa into the submucosa, it began to irritate the nerves, causing pain. If it reached the point of destroying the blood vessels in the submucosa, then a **bleeding ulcer** would be created. Normally, this just causes pain and some blood in the stool. However, as with the patient in the opening paragraph, occasionally a major **artery** can be damaged, causing massive bleeding, blood exiting from the mouth, and rapid death. Finally, after long periods of time, it was thought that the acid would eat through all the tissue layers, creating a **perforated ulcer**. The standard treatment for ulcers, therefore, was to ask the patients to reduce their stress level, to watch what they ate, and to take both antacids to lower the acidity level in the stomach and coating agents to protect the ulcer from the action of the gastric juices.

The problem was that we were never able to really "cure" this type of ulcer; we could only treat the symptoms. In the early 1980s, two Australian physicians, Barry Marshall and Robin Warren, noticed that many of their ulcer patients also had a high level of a bacterium called *Helicobacter pylori* in the mucous layer of the stomach. Further testing supported their idea that most ulcers were actually caused by an infection of the mucosa layer due to *H. pylori* that was then aggravated by the stomach acidity. Apparently, the bacteria release an enzyme that is capable of destroying the protective mucous layer, allowing the bacteria to cause a localized infection.

Unfortunately, even doctors and scientists have a hard time changing old ideas, and it took another 12 years before national and international medical organizations began to support this new idea about an old problem. Therefore, studies over the next several years showed that many medical personnel and most ulcer patients still believed that stress and diet were the cause of peptic ulcers. More importantly, this meant that less than 5% of ulcer patients were being treated with **antibiotics** for their ulcer infections, a proven way to effectively cure ulcers. It is only in the last few years that the medical community has come to widely accept the true cause of ulcers, allowing us to finally cure ulcers for the first time.

Such ulcers, however, are not restricted to the stomach, as the gastric juices flow out of the stomach into the **small intestine** with the food as it passes into the next organ in the digestive system. Since the small intestine is constructed very similarly to the stomach, the gastric juices and the *H. pylori* can cause the very same problem in the first section of the small intestine called the **duodenum**. Fortunately, while the intestine is called "small," this refers to its diameter, not its length. In fact, this intestine is approximately 20 ft. long, which allows the latter portion of the intestines, called the **jejunum**, to have relatively few problems with ulcers. This is due to the increase in pH and the lack of *H. pylori* the further away from the stomach you go. One significant difference in the anatomy of the small intestine is the presence of **villi** and **microvilli** lining the mucosa layer. These microscopic finger-like projections greatly increase the surface area of the small intestine, allowing increased **absorption** of the food into the blood vessels found in the submucosa layer. This process occurs only after **chemical digestion** has been completed, most of which also occurs in the small intestine.

Any food that cannot be digested or is not absorbed, moves from the small intestine into the **large intestine**. Again, the term "large" refers only to the diameter of the intestine, as this is about 5-6 ft. in length. Right where the small and large intestines join is the **appendix**, which was once thought to be a useless **vestigial organ**, no longer of any use to humans. We now understand that this is still an important structure that plays a role in preventing infection of the small intestine by bacteria found in the large intestine. The intestine then goes up along the right side of the abdominal cavity, the ascending colon, before crossing over under the stomach, the transverse colon, and coming back down along the left side of the cavity as the descending colon. The small intestine is coiled up inside the arch formed by these three sections of the large intestine. The large intestine ends in the **rectum**, which exits the body at the **anus**.

The presence of large numbers of bacteria in the large intestine creates, basically, an anaerobic fermentation tank where undigested and unabsorbed foods are further broken down and processed for **elimination**. While some necessary nutrients are released by this process and can be absorbed into the lining of the large intestine, most of the material here is destined to be released as **feces** in a process known as **defecation**. Fortunately, these bacteria are not likely to attack the lining of the intestines as *H. pylori* does in the stomach and small intestine. Therefore, ulcers are rarely seen in this organ.

Questions about this case:

1. What is peristalsis?
2. What is the common term for the process of "reverse peristalsis"? Why does it occur?
3. Not all bacteria in the digestive tract are harmful. What is the commonly known benefit of having some *E. coli* in your intestines?
4. What is acid reflux disease?

Questions to go deeper:

1. According to the case, the large intestine doesn't seem to do much. What is the function of the large intestine?
2. Why do you think it takes scientists and physicians so long to change their practices?
3. The stomach secretes acid to help with chemical digestion. What are some of the enzymes that help in this process and where are they produced?

References:

http://www.cdc.gov/ulcer/history.htm
http://hed2.bupa.co.uk/fact_sheets/html/Peptic_ulcer.html

Imagine Ten Boeing 747s Crashing Each Day

A Case Study of Diarrhea and the Digestive System

While Airbus has a larger model undergoing testing, currently Boeing 747 is the largest commercial airliner flying. It has several variations, but can carry up to 500 passengers. On May 25, 2002, one of these airliners crashed, killing all 225 passengers and crew. The world was horrified, as you can imagine. Newspapers, television, and radios carried extensive coverage of this disaster. Weekly newsmagazines continued with in-depth coverage for weeks after the accident. So why when we lose the equivalent of TEN Boeing 747's A DAY to diarrhea do those same media sources not say a word? Throughout the world, the leading cause of death is dehydration due to diarrhea. Diarrhea is a common symptom of a variety of illnesses with numerous solutions. Yet AIDS, bird flu, cancer, and other diseases get all the press. Obviously, the very mundane topic of diarrhea is not very newsworthy. To understand how such a common symptom, which we all have experienced, can cause such devastating effects, it is necessary to understand the physiology, or functioning, of the digestive system.

The functions of the digestive system are usually expressed as **mastication**, **digestion**, **absorption**, and **elimination**. Mastication is commonly called chewing, and obviously occurs in the mouth. It is one of the processes involved in the physical digestion of the food we eat. Another form of physical digestion occurs in the **stomach**, as the churning of the food and **peptic juices** occurs due to the **smooth muscle** in the lining of the stomach.

The process of physical digestion serves to increase the surface area of the food so that chemical digestion can occur. It also serves to mix the food with the various digestive enzymes being released as the food moves through the organs of the digestive system. Chemical digestion occurs throughout the length of the system, beginning in the mouth. There, **saliva** contains **amylase**, the enzyme required to begin the breakdown of starch. As you chew your food, mix it with your tongue, and swallow it, the amylase begins breaking down the starch into a sugar called maltose. This process continues as the food moves down through the **esophagus** and into the stomach.

The epithelial cells lining the mucosa layer of the stomach secrete **gastric juices** that contain **HCl** and **pepsin**. As the stomach churns the food together with the juices, the pepsin starts to act on the **protein** in the food, breaking it down into **polypeptides**. However, this will only occur under acidic conditions provided by the presence of the HCl, commonly known as hydrochloric acid. If either the HCl or the pepsin is not present, this process will not occur.

Secretory cells are found throughout the digestive system and are also very important in the **small** and **large intestines**. They are also the cells that when affected lead to diarrhea, the subject of this case study. In the small intestine, these cells release enzymes that are involved in the chemical digestion of food. In addition to the enzymes released by the small intestine, there are also enzymes that are released by other organs that have their action in the small intestines. These organs are called **accessory organs** and include the **liver** and the **pancreas**. The liver secretes **bile**, which is not really an enzyme. In truth it is an **emulsifier**, which means that it is going to break fat down into smaller fat droplets. This is the type of chemical that is found in Italian salad dressings to help make the oil, vinegar, and water mix. If you have ever tried to use one of those "natural" salad dressings, which lacks emulsifiers, you know how quickly the ingredients separate when you stop shaking the bottle. When the fat is broken down into smaller fat droplets, the surface area is increased, which increases the ability of the next enzyme, **lipase**, to break down the fat.

The enzyme lipase is released by another accessory organ called the **pancreas**. As it travels into the small intestines, it begins to work on the fat droplets, breaking the fat down into its smaller chemical constituents, **glycerol** and **fatty acids**. The pancreas also releases additional amylase, which works on any remaining starch, not digested by the salivary amylase, to break it down into maltose.

As mentioned previously, the small intestine also releases its own enzymes, which complete the chemical digestion of the major food groups. **Maltase** attacks the maltose, breaking it down into individual **glucose** molecules, which are small enough to be absorbed by the body. To complete the digestion of polypeptides, the cells lining the small intestine also release a variety of peptidases. The enzymes cause the polypeptides to be broken down into their constituent **amino acids**, thus completing the digestion of the proteins.

Besides serving as the primary location of chemical digestion in the digestive system, the small intestine also functions in the process of absorption. Now that the various organic compounds have been digested into compounds small enough to cross the cell membranes, they can be taken across the cells of the small intestine and moved into the **blood vessels** and **lacteals** lying behind the epithelial layer. The blood vessels are responsible for absorbing the water-soluble substances resulting from the digestion of proteins and starch. The material created when the fats are digested, however, are not water soluble, so they cannot go immediately into the blood. Instead, the glycerol and fatty acids are first absorbed into the **lymph fluid** inside the lacteals. These structures are part of the **lymph system** and will process the fatty material until it can be dumped safely into the circulatory system.

However, since these cells are involved in the absorption of material, they can also be involved in the release of other material, and this is what happens in a disease known as **cholera**. This disease has been known since the 1500s and was originally found in India and surrounding countries. It then spread into Europe and North and South America in the 1800s. However, more recently, the world has been dealing with a worldwide **epidemic**, known as a **pandemic**, as the disease has spread from Southeast Asia to India to Northern Africa and, most recently, to South America. In fact, there have also been cases occurring in the Western United States due to vacationers returning from South America to Los Angeles.

This disease is caused by the bacterium known as ***Vibrio cholera***, which is transmitted, it is thought, in the feces of patients with cholera, which gets into the water supply. From there it can be transmitted to another human if he or she drinks the contaminated water or eats certain types of organisms living in the contaminated water. Once the bacterium is in the human, it releases a special kind of **toxin** that affects the cells lining the small intestine. The toxin causes cells that line the small intestine to stop absorbing food molecules and begin pumping water into the lumen of the intestine. Along with the water, they also pump out important **electrolytes** from the blood and tissues behind the epithelial cells. The pumping out of the water is what causes the principle symptom of cholera—massive diarrhea, commonly called rice-water stool. So much water can be released that patients can die within 2 to 3 hours, although death is more likely to occur after several days. The loss of electrolytes, which are required for a number of important functions such as nerve function and muscle movement, can lead to heart failure and circulatory failure.

A more common illness that also affects the lining of the small intestine and may have deadly results is caused by *Escherichia coli*. Though it is a common inhabitant of the human digestive system, if you encounter too many of the bacteria or a **strain** your body is not used to, you will develop **traveler's diarrhea**. While this is usually more likely to be an annoyance than a disaster on your overseas trip, it is the leading cause of infant mortality in the world. Even in our country, deadly strains of *E. coli* have killed children who have eaten, usually, undercooked hamburger meat. As with cholera, the bacteria do not invade the lining of the intestine or cause inflammation. However, the toxins they release cause the epithelial cells to excrete large quantities of liquid.

Diarrhea can also occur if something affects the cells lining the **large intestine**. Normally, this organ is the site of **elimination**, the final function of the digestive system, along with some **water absorption** and release. As part of the digestive process, large quantities of water are added to the food and enzymes to enhance mixing and digestion. If the cells in the large intestine are functioning properly, they will absorb the excess water, creating feces with the proper consistency for elimination. When they do not work the way they should because of viral, bacterial, or other infections, they release water instead of absorbing it and, again, diarrhea is the result. Two of the most common forms of food poisoning have their principle action in this way. **Salmonella food poisoning** is due to bacteria in the *Salmonella* genera. Salmonella food poisoning is on the rise in the United States primarily because of our country's increased reliance on poultry products. As we have learned more about the potentially harmful effects of too much beef, we have been eating more chicken and turkey. Unfortunately, these two organisms tend to have high levels of harmful *Salmonella* bacteria due to the way modern poultry is raised. This allows the meat, eggs, and anything made from them to be potential transmitters of the bacteria. Once in the large intestine, these bacteria invade the lining of the colon causing an **inflammatory response** that affects the permeability of the blood vessels in the area. This invasion of the lining of the intestine causes diarrhea that can be quite severe in some cases.

Most forms of diarrhea that cause significant problems around the world are due to **pathogens** transmitted through water contaminated by human feces. This means that to prevent these diseases, the water either has to be uncontaminated in the first place or needs to be cleaned up before it is drunk. If this is not done, and patients begin to suffer from diarrhea, then they must be **rehydrated**. Unfortunately, rehydration means giving the patient water to replace the liquids lost through diarrhea. If the initial problem was due to the contaminated water in the first place, trying to rehydrate with the same contaminated water will only aggravate the problem. This is why one of the most helpful things people can do to save lives around the world is to teach what we would consider common sense. First, we would need to teach people to go to the bathroom away from where they get their water. Then we need to teach them to filter or otherwise treat the water before they drink it. We would also have to teach people to wash food in water that is not contaminated. Along with this, however, are some preventative measures that require a greater investment, such as digging wells, building sewage systems, and preventing wastewater from contaminating rivers and streams. Finally, teaching people how to properly rehydrate patients is critical to caring for those suffering from diarrhea. The use of boiled water and the addition of common salts can mean the difference between life and death from what, for us, seems like simply a nuisance symptom.

Questions about this case:

1. Not all bacteria that infect our intestines are harmful. What are the names of some beneficial bacteria in our intestines and their functions? What effect do common antibiotics have on these bacteria?
2. If a friend was visiting Mexico or India and wanted to avoid getting diarrhea, what advice would you give him or her?
3. Why do you think the leading cause of death worldwide gets so little publicity?

Questions to go deeper:

1. What is the role of the gall bladder in the process of digestion?
2. What is acid reflux disease and what are some of the common treatments?
3. How do enzymes that are made in the pancreas get to the small intestine?

Silent Spring in Fort Lauderdale

A Case Study of Human Impact on the Environment

Growing up in Fort Lauderdale, Florida, was a great experience. Known as the "American Venice" due to its extensive network of canals that connect most of the city to the Intracoastal Waterway and the Atlantic Ocean, it was a life of sailing, skiing, fishing, and just about anything else that was water related. It was also the hot spot for spring break, and thousands of college kids converged on the city during their break. While this was an economical boon for the city, it was anything but a quiet time, with traffic jams, outdoor concerts, all-night parties, and dozens of other noise-producing events occurring. Several years things got a little noisy and out of hand and, after the police had calmed things down, civic leaders and residents would call for the halt to the activities. However, in the 1960s, Fort Lauderdale was suffering from much more important, and more subtle, problems.

Like much of the United States, there was trouble in paradise. Trash littered the waterways, raw sewage was dumped directly into the ocean, fish levels were dropping, and *Pelecanus occidentalis* was disappearing from the shores. *Pele* what? Commonly known as the brown pelican, this large, ungainly bird was a frequent inhabitant of the coastal waters on the Atlantic, Pacific, and Gulf Coasts of the United States. Soaring majestically just a few feet off the water's surface, swooping down from a height and plunging into the water with beaks agape, these birds came to symbolize the beauty of Fort Lauderdale for me.

However, I remember wondering one day when I had last seen a pelican. It was one of those things like watching a young child grow. If you are with them all the time, you don't really notice that they are slowly growing. But if you go away to college for a few years and then come back, you immediately notice how much they have grown. My experience with the pelicans was similar in that they were there, and then they weren't. It took a while for us to notice that what had always been there was gone.

I was devastated: first, because I loved to watch them; second, because I knew their absence meant something was happening; and third, because I could not believe they had disappeared and I hadn't noticed it. Apparently, I wasn't the only one who fit into these three categories, as biologists throughout the area had finally awakened to the disappearance of these creatures. By the end of the 1960s, these once numerous birds had virtually disappeared from the Gulf Coast, West Coast, and most of the East Coast. In fact, a few islands on the Gulf Coast of Florida were about the only places they continued to nest and, by 1970, they had been placed on the list of endangered species.

What had happened? The initial suspect was a chemical known as **DDT**, which was being used extensively as an insecticide throughout most of the world. As Rachel Carson had reported in her 1962 book, *Silent Spring*, this and many other insecticides were having a dramatic but, at the time, unnoticed effect on many species of birds. When she wrote what was soon to become an environmental classic, the thought was that these insecticides were having primarily an indirect effect on many birds because, as they were used to kill the "bad" insects, they were also killing the "good" insects that many birds relied upon for food. There was also some evidence that these chemicals were directly killing the birds when the insecticides were sprayed from trucks, airplanes, etc. Thus, instead of the delightful springtime songs of the wrens, finches, and other songbirds, Americans were waking up to a silent spring as many of these species died.

DDT was also blamed on the death of birds like the pelicans and other predatory birds like eagles and ospreys because of a process known as **bioaccumulation** or **biomagnification**. In this case, the birds were not being killed due to direct exposure to DDT but due to the increased concentration of the pesticide that was occurring along the **food chain**. A food chain is a series of "who eats who" situations. The first member of the series is a **producer**, which is any organism that is capable of doing a process like **photosynthesis**. Producers can sit in the sun, absorb water from the soil and carbon dioxide from the air, and turn it all into plant matter. These are eaten by a **primary consumer**, what is commonly called an **herbivore**. These creatures eat a lot of producers in order to make more animal matter. Anything that eats a primary consumer is a **secondary consumer**, which we commonly call a **carnivore**. However, there are also animals that we know as **omnivores**, which can occupy virtually any position in the chain because they are capable of eating both producers and other consumers. The creatures that eat secondary consumers are called **tertiary consumers,** and the list goes on.

Whenever one organism eats another organism, energy and material are transferred from one to the other. Unfortunately, no such transfer of energy is 100% efficient. Instead, the majority of the energy is "wasted" as heat or waste products that the organism cannot absorb or make use of. A substantial amount of the energy is also used to maintain the organism in its daily life. It takes a great deal of energy to simply keep the organism alive, to repair damaged tissue, or to replace cells lost due to wear and tear. This leaves very little energy for the construction of new cells required for the organism to grow. This means that each level of the food chain must eat more and more of the preceding level in order to stay alive.

This appears to have been what was occurring when scientists studied the level of DDT in one particular ecosystem. The level of DDT was found to be very low in the water itself, in the order of 0.00005 ppm (**parts per million**), a seemingly insignificant amount that could not possibly, it was thought, harm birds or humans. However, upon further evaluation, it was found that DDT had a tendency to accumulate in the tissue of organisms and that its amount was magnified because, as you move along a food chain, each successive organism eats more and more of the smaller organisms in the chain. In the case of DDT, the effect is enhanced by the fact that DDT remains in the fatty tissue of organisms and accumulates over the years. Thus, the plankton that lived in the water were found to have DDT at concentrations of 0.04 ppm. The shrimp that lived in the water and ate the plankton had levels around 0.16 ppm, and the eels that ate the shrimp had increased levels of DDT near 0.28 ppm. The larger fish that ate the eels and the shrimp had accumulated levels of 2.07 ppm and, in this particular study, the sea gulls that ate the fish had DDT levels of 75 ppm in their fatty tissue.

This clearly showed that while the levels of toxins in the water could be insignificantly low, they could still cause problems in higher levels due to the accumulating and magnifying effect that occurred. However, even these levels were shown to not be directly poisonous to the birds in most cases. So why were the birds disappearing? While the answer is still under debate, the most common understanding is that the high level of DDT in the predatory birds was affecting the bird's ability to metabolize **calcium** properly. Since calcium is a necessary ingredient for strong egg shells, the lack of calcium was causing the female birds to lay eggs with thin egg shells that cracked, killing the chicks, just due to the weight of the mother sitting on the eggs. When birds like the brown pelican lay only two to three eggs a year, the loss of even a few eggs each year was having a dramatic effect on the bird population.

Further study since the 1960s has shown that, while this seemed to explain what was happening with osprey and eagle populations, the situation was even more complicated for the brown pelican, especially in Florida. The pelican was suffering from what may be called an ecological complex involving a number of problems that conspired to reduce their populations. Besides DDT, another pesticide called Endrin was also being dumped into the water supply. This pesticide was causing large-scale fish kills of the smaller fish that the pelicans usually feed on. Other forms of pollution were also contributing to reduced fish populations, as was the destruction of coastal areas due to development, excess runoff from farms and cities, rising sea levels, and a number of other factors. Thus, the pelicans were disappearing due to a simple lack of food.

These birds usually nest on low-lying coastal areas and islands lying just off the coast. Unfortunately, *Homo sapiens*, commonly known as humans, desiring a beautiful waterfront home, also prize these areas. During the late 1960s, these areas of Florida, particularly on the East Coast, were being developed at an alarming rate, with huge condominiums and hotels being built right on the beach, while million-dollar homes were going up along the canals and waterways. Even spoil islands–islands that have been created from the material dredged up while creating and maintaining the waterways–were being converted into developments, instead of being allowed to remain as nesting sites for the pelicans. Thus, these creatures were being hit by a double or triple whammy of increased pollutants, decreased food supplies, and disappearing habitats.

Thanks to Rachel Carson's book and the uproar that ensued, the U.S. government began to take action. In 1972, DDT was banned for use in the United States, and Endrin was removed from use in 1984. These actions and others also increased the population of the fish pelicans feed on. Finally, the establishment of regulations protecting coastal habitat for nesting sites and the birds' own adaptation to new sites allowed the return of these birds to the coastal waters of America. While this occurred naturally, it was aided by extensive efforts by various public and private groups who worked to reintroduce pelicans from Florida into areas completely depleted of the birds, like Louisiana, Texas, and parts of California. These efforts resulted in the birds along the East Coast and in Florida being taken off the list of endangered species in 1985 and, hopefully, will allow them to be taken off the list throughout their American range.

Since the populations of sea gulls, ospreys, and eagles have also begun increasing with these changes, it is obvious that, just as humans can mess up the environment, they can, in some cases, repair that damage also. Thanks to courageous individuals like Rachel Carson, this situation has been corrected; unfortunately, it would have been significantly easier, cheaper, and simpler to prevent the situation from occurring in the first place. Now, if only something could be done about those bothersome college students.

Questions about this case:

1. The study mentions the effect that killing the good insects had on songbirds. What would be another significant impact on the environment of losing so many insects?
2. Pelicans are an example of a type of organism known as a **top consumer**. Why?
3. Pelicans are not the birds that were most affected by DDT, nor were they the ones who received most of the press regarding the effects of DDT. What other birds are more famous for having been affected by DDT?

Questions to go deeper:

1. Food chains are nice, but not very realistic. Why?
2. If food chains are not the way things are in the real world, what is the model we use called?

References:

http://www.nwf.org/nationalwildlife/article.cfm?articleid+689"&issueid+27

http://www.chem.duke.edu/~jds/cruise_chem/pest/effects.html

The Curse of the Centennial Expositions

A typical Southern forest overgrown with kudzu.

A Case Study of Ecology and Why We Should Leave Well Enough Alone

The Philadelphia Centennial Exposition of 1876 and the 1884 World Industrial and Cotton Centennial Exposition in Louisiana were highlights of their time. The latest in "modern" technology, techniques, and whatever were on display. People came from all over to witness the amazing things being shown. They then took home with them the sights, sounds, and, in some cases, plants being displayed—and that was the problem.

From the Philadelphia Centennial Exposition people took samples of a Japanese plant known as **kudzu (*Pueraria montana* var. *lobata*)**, a beautiful ornamental vine with large, fragrant purple flowers whose large leaves provided a tremendous amount of shade. This amazing plant found a perfect growing environment in the southern United States, where it is capable of growing a foot a day and 60 feet a growing season. This made it ideal to help control erosion and to provide forage for some types of livestock.

Just a few years later, a South American water plant called **water hyacinth (*Eichhornia crassipes*)** was introduced at the World Industrial and Cotton Centennial Exposition. Again, the beautiful, flowering plant was seen as a lovely ornamental, due to its large leaves and showy flowers. A Florida man apparently took home some samples of this plant and introduced it into the St. John's River in northern Florida thinking it would make a lovely addition to the waterways of Florida.

In both these cases, the people involved were sadly mistaken about the usefulness of these nonnative plants, as they soon became a textbook example of well-meaning folks causing major environmental problems. In the case of kudzu, the plant was spread throughout the South by the Soil Conservation Service, either through the use of the Civilian Conservation Corps or by paying farmers to plant kudzu in an effort to control erosion. By 1946, some three million acres had been planted; unfortunately, by 1955, the plant had moved outside of its original plantings, until it now covers some seven million acres.

It now covers not only eroded lands, but also valuable forest and cropland. Here it covers trees and plants, cutting off the sunlight they need to grow, weighing them down under the tremendous weight of its vines, and covering all manner of human artifact. Driving along most southern highways, one passes mile after mile of vine-covered landscape. And still it grows, as a method of control has yet to be found. Most herbicides will not bother it in the least, and some will actually help it grow faster. A few seem to be effective if given over long periods of time but even they may require some 10 years to work.

Some researchers have found that if the vine is grazed regularly, its growth can be controlled, but few animals are capable of consuming the tough weed. Angora goats can survive on kudzu while providing milk and wool, but having these goats grow on seven million acres is impractical. Thus, a problem started over a 100 years ago continues to plague the South.

The South continues to suffer from those ornamental water plants someone thought would make an attractive addition to Florida waterways. Water hyacinth is one of the most productive plants in the world, with a doubling time of about 12 days. Their primary method of reproduction is by **asexual budding** and **stolons**, which form daughter plants at a prodigious rate. These plants create dense mats that currently cover some 125,000 acres of water in Florida alone. They can also reproduce **sexually** via **seeds,** as do most flower- and cone-bearing plants. Since these seeds can lie dormant for up to 20 years, waiting for the proper conditions, they can produce new infestations long after the original plants have died.

Just like kudzu, these plants can cause both environmental and economical damage. Their rapid growth can cover the surface of the water with up to 200 tons of plant material per acre, blocking out sunlight, crowding out other plants, and creating a virtual **monoculture** of water hyacinths. As these plants die, they begin rotting, which quickly creates an **anaerobic** environment, as oxygen is used up in decomposition. Even more plants and animals in the water die as this happens. The thick mats also block waterways, stopping boat traffic, fishing, and swimming in the areas affected.

Again, as with kudzu, state and federal governments have spent millions trying to control the growth of this aquatic weed. The plant is susceptible to biocides, and all manners of strange machines have been developed to remove the plants from infected waterways. In addition, **biological controls** have been found, including insects that attack the water hyacinth. Now, according to the state of Florida, the plant is under maintenance control, which means that as long as the various maintenance techniques are in place, the plant is pretty much under control. However, if that original plant was never brought into the United States, or if it had remained in the display and not been introduced into Florida's waterways, all of this would have been unnecessary.

If these plants do not belong here, how are they able to do so well and how can they replace native plants that have lived there for, in many cases, centuries? The answer lies primarily in the concept of the **niche**, which is commonly referred to as the organism's role in the environment. However, this role must be described in relation to the various organisms it interacts with in one way or another. Plants occupy the generic niche of being **producers**, which means they are the organisms that convert **inorganic** material like sunlight, carbon dioxide, and water into **organic** matter in the form of new plant material. However, each specific organism has a much more detailed niche that it fills in its particular **ecosystem.** One plant's niche might be to grow in areas with exposed soil that lie in direct sunlight, while another's might be to grow in the heavy **litter** and shade produced by other plants.

When new species invade an area, they may find unoccupied niches just waiting for them. If no other native species is filling that role in the environment, they can just move in and become established. In other cases, an invading plant finds all the niches filled and can only move in if it can **compete** successfully with the native plants that are already there. This means that it must be able to do something better than the other plants. It might be able to grow with a little less water or a little less nitrogen in the soil. It might be more resistant to a plant disease that is common in the area, or it may produce a prettier flower that is better at attracting animals to pollinate the flowers or disseminate the seeds. Whatever its advantage, if it can outcompete the other plants, it will eventually replace them.

Plants are not the only creatures to have been introduced into areas where they did not belong with devastating consequences. Numerous birds, insects, marine animals, and rodents have been purposefully or accidentally introduced throughout the world. As you may know, the British colonized Australia primarily with convicts. However, quite a few early settlers were of a higher class of citizens, and one such man had enjoyed hunting rabbits in the English meadows and forests. In 1860, he asked a friend to send him a breeding population of rabbits and, as rabbits do, they reproduced quite rapidly. He released many into the wild and gave away countless others to friends interested in having some sport. Unfortunately, these rapidly procreating creatures quickly out-ate, outcompeted, and out-reproduced the native populations of small grazers. They literally ate them out of house and home, as they even took over the burrows of the native inhabitants. Eventually, the rabbit population got so out of hand that they destroyed over four million square kilometers of Australia. Besides the ecological impact of this, there was also a tremendous impact on the economy of Australia. Not only were sheep and cattle ranchers at a loss due to the devastation of the grazing lands, but the entire country suffered from the cost of building hundreds of miles of fences, trying to replant denuded areas, damage from erosion, and the attempts to control the population via hunting, trapping, and poisoning. While large parts of the country are still suffering from the rabbit influx, the introduction of a virus that is deadly to rabbits has worked well enough that, in much of the country, the ecosystem is recovering from the effects of the Englishman who wanted to hunt.

An example of an invasion that was accidentally accomplished involves the **zebra mussels** who naturally inhabit the Caspian Sea in Central Europe. Several decades ago, they started arriving in the Great Lakes area, initially around Detroit. They were apparently carried in on freighters coming from Europe to the United States, and quickly found a suitable habitat in the freshwater of this area. While smaller than the typical marine mussels you might see growing on a dock or a rock at the seashore, these are prolific reproducers, and their small size allows them to go where other similar creatures cannot. Boats, docks, piers, and other human devices were quickly covered by the mussels, requiring great efforts to remove them. Pipes and grates covering the pipes were quickly blocked, preventing the water from moving through them, which greatly impacted those industries that brought in large quantities of water to cool their machinery and processes. Water treatment plants that either take in or dump out water were also affected, as their pipes are blocked. Native species of freshwater organisms are also affected as the zebra mussels could grow on top of native mussels and outcompete them, leading to the disappearance of the natives. Since the introduced mussels live off the plankton floating in the water, and as they are so good at doing this and there are so many of them, they can lower the population of native fish that usually survive on the plankton. Unfortunately, all efforts to eliminate or even restrict the growth and spread of these mussels have failed, and they are slowly expanding their range in Canada and the United States.

Humans have also caused innumerable problems, by doing just the opposite; instead of adding creatures, they have removed organisms from certain environments. This has usually occurred with **predator–prey** relationships. In these cases, the predator is the one doing the eating, and the prey is the one being eaten. Cattlemen and other livestock men have often tried to eradicate predators in their areas believing that they are causing too much loss to their herds through **predation**. By hunting, trapping, and poisoning, they have killed the wolves and coyotes they blamed for killing their livestock. However, in many cases, the loss of the natural predators has led to astronomical increases in the natural **primary consumers**, more commonly known as **herbivores**. Normally, the primary consumers are kept under control by their natural predators. As these predators disappear, the grazers overpopulate the area, and they begin outcompeting the livestock for the grass and such, leading to even more loss as the livestock run out of plants to eat.

Many other examples of the ecological and economic impact that introduced species have are, unfortunately, too easy to find. Whether the species are deliberately introduced by humans for what, at least initially, seems like good reasons, or they are accidentally or even naturally introduced really doesn't matter. The impact is almost always detrimental and ends up costing millions of dollars, while causing, sometimes, irreparable damage. How much easier it would be if we simply followed the age-old adage that "It isn't nice to fool (with) Mother Nature!"

Questions about the case:

1. What are the different components of a species' niche?
2. Describe another example of purposeful or accidental introduction of a species to a new area that had detrimental effects.
3. Did you notice that one way to help control water hyacinth was to introduce an insect that attacks the plant. What could be the potential problem of that? What are some good questions to ask before this insect is introduced that might help limit these problems?

Questions to go deeper:

1. I am not sure what might surprise you most: that plants can reproduce sexually or asexually. Describe how these differ.
2. The case described a potential problem that a huge plant mass (of something like water hyacinth) could create if introduced in an aquatic environment. What would be the chain reaction of problems?

The Goldilocks Principle

A Case Study on the Complex Issue of Global Warming

Remember when Mom and Dad read fairy tales to you? Remember the story of "Goldilocks and the Three Bears"? Well, in this case, we will be looking at a variation on that, called "Goldilocks and the Three Planets." When astronomers and climatologists refer to the Goldilocks Principle, they really are referring to the old fairy tale. If your memory isn't failing you, you know that when Goldilocks was invading the bears' house, she kept running into three of everything—three chairs, three bowls of porridge, and three beds. In each case, the poor girl finds out that one is too something, the other is too something, and the third one is just right.

Well, in this case, we are talking about three planets, Venus, Earth, and Mars. We are interested in the temperature of the planets and how that affects life on them. When it comes to temperature, Goldilocks would find Venus too hot, Mars too cold, and Earth just right. The issue at hand is why this is true and how it relates to **global warming**. Well, the answer to the question involves several aspects including the size of the planets and their distance from the sun. However, the issue of global warming is also affected by a concept called the **greenhouse effect**.

Now, there is a lot of confusion, disagreement, and uncertainty about both of these important concepts, but the general understanding is as follows. The planet Venus is closer to the Sun and somewhat smaller than Earth. It has an atmosphere that is 90 times denser than Earth's and contains 96% **carbon dioxide** (CO_2). Because of this, its temperatures are too hot, around 460 °C (860 °F), which is hot enough to melt lead! Mars, however, has an atmosphere that is one-hundredth as dense as Earth's and, while the CO_2 concentration is still high, making up 95% of the Martian atmosphere, there is so little of it that the temperatures are too cold, varying from –113 °C (–171 °F) to a balmy 0 °C (32 °F). Sandwiched between these two uninhabitable planets is good old Earth. With an atmosphere whose density is smack in between Venus and Mars, and with a CO_2 concentration of 0.036%, we have an average global temperature around 15 °C (60 °F), just right for us human types, as Goldilocks would say.

What does the amount of CO_2 in the atmosphere have to do with the temperature of these planets? CO_2 is one of the **greenhouse gases**, which includes water vapor, methane, nitrous oxide, and ozone. This group of gases shares the ability to allow sunlight, which exists in short wavelength energy, to pass through them with little effect. However, when this short wavelength energy hits something in the atmosphere or hits the earth's surface or anything on the surface, it is absorbed and reradiated as a form of energy that has a longer wavelength. This form of energy cannot pass through the gases and, instead, is absorbed by the material in the atmosphere so that a lot of the heat is retained in the atmosphere. This means that the atmosphere gets warmer because it is being heated by two sources: the Sun and the radiated heat from the Earth.

In the case of Venus, the dense atmosphere contains so much CO_2 that virtually all of the heat is retained, which, over time, has created extremely high temperatures. On Mars, the lack of CO_2 in the atmosphere means that virtually no heat is trapped and very cold temperatures are the result. Only on planet Earth is the density of the atmosphere and the amount of CO_2 just right for creating temperatures that life, as we know it, needs. So, could the amount of greenhouse gases in Earth's atmosphere ever get high enough to create Venetian temperatures on Earth? That is the big question that is causing so much debate.

The first question is whether the amount of CO_2 in our atmosphere is really increasing, and the answer seems to be a resounding yes. Scientists have been accurately measuring the amount of CO_2 in the atmosphere since 1958, when the level was recorded as 316 ppm. In 2004, that level had risen to 377 ppm. Scientists believe that they have also been able to accurately determine levels of CO_2 back millions of years, using ice cores from glaciers, ocean sediments, tree ring growth, and other techniques. While these studies conclude that CO_2 levels have definitely risen since the Industrial Revolution, when levels were around 275 ppm, it is obvious that significant and consistent cycles have existed for the last 500,000 years. In these cycles, levels have fallen to below 200 ppm and risen to over 300 ppm. Going back even further, studies show that 150 million years ago, when dinosaurs dominated the landscape, CO_2 levels were probably ten times higher than they are now. Therefore, while levels are higher now than they have been in several hundred years, they are not higher than they ever have been before.

The greatest amount of controversy and disagreement, therefore, seems to be about the cause of the increased CO_2 levels. This is the second question at hand. Obviously, if CO_2 levels have varied widely and if levels have approached or exceeded today's levels in the past, before humans produced massive amounts of CO_2 through the combustion of fossil fuels, something else must have accounted for the high levels of CO_2 besides human activities. Environmentalists believe that currently humans are producing excess amounts of CO_2 by burning abundant amounts of **fossil fuels**. Since these fossil fuels, by definition, were formed from ancient fossilized tropical rain forests, we are essentially burning carbon that was trapped in plant matter millions of years ago. As that carbon is released, it forms CO_2, which can undergo several fates. Much of it, apparently, enters the atmosphere, where it is apparently causing the atmospheric concentration to increase. However, much of it is also absorbed into the world's oceans, where it is dissolved and may become calcium carbonate, which is used by marine organisms to make their shells. Finally, a large amount of it is absorbed by vegetation, which uses it in the process of **photosynthesis** to create new plant matter.

While the first fate is what is believed to contribute to the increased CO_2 in the atmosphere, the last two mechanisms actually remove CO_2 from the atmosphere. Therefore, anything we can do to prevent the creation and release of CO_2 or to increase the amount of existing CO_2 that is absorbed by the oceans or vegetation should decrease the amount of CO_2 in the Earth's atmosphere. Thus, we have environmentalists calling for decreased burning of fossil fuels in **internal combustion** engines, the principle producer of CO_2, and for decreased cutting of forests, especially those in the tropics, which can absorb large amounts of CO_2 from the atmosphere. It is hoped that if these two events occur, then the levels of CO_2 will, at best, decrease or, at worst, remain level.

The final question that needs to be addressed is whether the higher CO_2 levels we now have are causing the Earth's temperatures to increase. At present, this is a hotly contested issue that seems to have as many answers as the number of researchers studying it. Several years ago, scientists who believed they had found a way to determine the Earth's temperature as far back as 1000 AD produced the now classic "hockey stick" graph. What they found showed a relatively flat temperature graph up until 1900, when the Earth's temperature suddenly began increasing sharply. This long flat period with a sudden, sharp upturn created the so-called "hockey stick" shape. However, sharp controversy has arisen over this graph. Detractors point to poor data analysis, the misleading omission of "**error bars**" showing the wide variation inherent in much of the data, and even the omission of critical past periods of cooling and heating during the time period under discussion. Therefore, while the majority of environmentalists seem to believe that **global warming** is a reality, there does seem to be a great deal of disagreement among climatologists as to some aspects of this.

Perhaps, then, the best way to approach this is to look at the possible consequences of an increase in CO_2 levels and the resultant increase in temperature. This way, if the possible consequences are not that great, we can see that we really do not have to worry about having an Earth that is getting hotter. However, if the possible consequences are significant and far-reaching, then perhaps we should start taking steps to prevent even a small amount of temperature increase. Fortunately, we are not, in anyone's mind, looking at an increase on the magnitude of what is occurring on Venus. However, even a slight increase in temperature could have significant implications for a variety of events. One that is talked about a lot is the impact on the level of the world's oceans. If global warming is occurring, it will take a very small amount of increase to significantly increase the level of the oceans. A small amount of increase would occur due to the melting of the glaciers and the polar ice caps. However, the majority of the increase would occur due to simple expansion. Remember that as any object increases in temperature, the molecules become more active and spread out, causing an expansion of the structure. Since the oceans cover so much of the Earth's surface and extend so deep, even a small amount of increase could cause a significant increase in the level of water. Currently, estimates are that sea levels are rising 2mm/year, which may not seem like much. However, since 75% of the world's population lives within 60 km of a coast and many countries in the South Pacific are only a few cms above sea level now, even a small increase in ocean level could be significant.

Another important consideration is the impact that changing global temperatures will have on the Earth's climate. While a discussion of the formation of our climate is beyond the scope of this study, it is based primarily upon **differential heating**, which occurs due to the spherical shape of the Earth, the Earth's tilt, and other factors. If the Earth begins to increase in temperature, this differential heating will change, causing global wind patterns to change and high and low pressure areas and rainy and dry areas to shift. If this happens, the Earth's desert areas and its rainy areas could change, impacting millions of humans. Areas that are now known for producing much of the food the world depends on could become too dry to produce crops. Hurricanes and monsoons could become more numerous and more severe. El Niños would become more severe and occur more frequently, changing ocean temperatures, storm and rainfall patterns, and even drought patterns. The very basic water cycle, more technically known as the **hydrologic cycle**, could also be altered. Some environmentalists believe that there is evidence that some of these things are happening already.

If these things begin to happen, life on earth will become unpleasant at best, life-threatening at worst. If these things do not happen, life will continue on pretty much as it always has. Therefore, in spite of the controversy and disagreement over whether all this is occurring, it seems logical to take a worst-case scenario and begin taking the steps necessary to reduce the amount of greenhouse gas emissions and to reduce the destruction of forests. If environmentalists are correct, we may very well save thousands of lives. If they are wrong, we will have simply initiated pollution regulations that will result in cleaner air, more efficient engines, and less reliance on fossil fuels, and we will have begun programs that will stop the destruction of important rainforests and all the organisms that rely on them. As Goldilocks would say—sounds just right!

Questions about this case:

1. Which has more energy, short wavelength energy or long wavelength energy? Is that related to the difference in how the atmosphere handles them?
2. If forest trees are good at absorbing CO_2 out of the atmosphere, would it be better to have small, young trees doing the absorbing or large, mature trees doing the absorbing? Why?
3. When internal combustion engines burn fossil fuels, they produce more than just CO_2. What other types of material are produced that cause environmental problems?
4. What do the error bars on the hockey-stick graph mean? Is this data reliable?

Questions to go deeper:

1. Carbon dioxide is not the only gas in our atmosphere. What are the other gases, in order of decreasing amounts?
2. Why might scientists be able to determine what the concentration of the old Earth atmosphere by examining cores taken from glaciers?
3. Most of the world's nations have banded together to sign the Kyoto Protocol. What is this and how does it relate to this case study?
4. Do you agree with the conclusion in the final paragraph? Why or why not?
5. Briefly outline the hydrologic cycle.

References:

http://www.crystalinks.com/greenhouseeffects.html

http://www.ems.psu.edu/~fraser/Bad/Bad/Greenhouse.html

http://www.ncdc.noaa.gov/oa/climate/globalwarming.html

http://www.ucar/edu/learn/1_3_1.htm

An Embarrassing Lesson to Learn

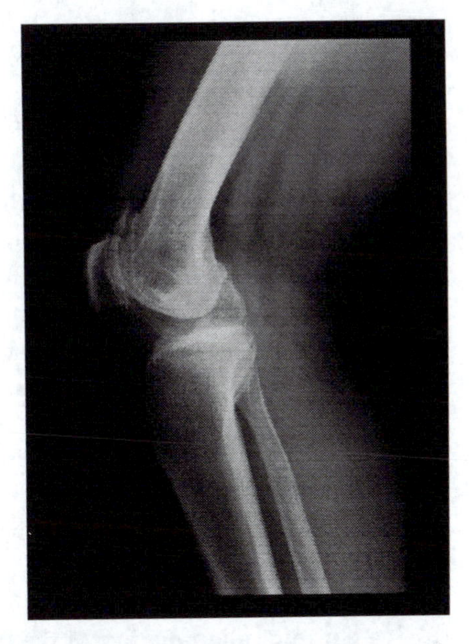

A Case Study of Rheumatoid Arthritis and Joint Anatomy

It was an embarrassing lesson to learn. Sometimes I would catch myself being a bit judgmental when I would see a person park in a handicapped parking space with the appropriate permit who would then walk into the store with no apparent difficulty. I would question in my mind how they got the privilege to park in that space without being seriously ill. Then my good friend Heather was diagnosed with **rheumatoid arthritis** (RA). I can remember driving with her to go shopping on a day when her pain was only moderately bad. I asked if she had her parking permit with her so I could hang it from my rear view mirror. She said she did, but said she did not want to use it. I vehemently protested. Today was a good day since she could function with moderate pain in her feet and hands, but in my opinion, she had every right to park close to the store. She was still in a good deal of pain, and walking only made it worse. Then she made the comment that pricked my guilty conscience. She said that on days when she could walk into stores without limping badly, she felt the stares of people when she parked in the handicapped parking space. It was as if they did not believe anything was wrong. I was stunned. I tried to think back to remember if I ever stared at anyone for doing that. It did not matter; I was guilty.

Heather was diagnosed with RA in her early 30s. She was an avid hiker, biker, and outdoors person. Over a period of time, she found that she had increasing joint pain. The pain would be worse the days after she was physically active, but she attributed it to simply getting older. It progressed to the point where the pain would just not go away. A trip to her doctor and then to a specialist determined that her problem was RA. It wasn't until Heather did some research on RA and how it typically progresses that she realized how devastating RA could be.

To understand what Heather is dealing with, we need to start with an understanding of joint anatomy and function. In terms of anatomy, a **joint,** or **articulation,** is any point in a body where two bones meet. It does not matter if the joint is movable like the knee, or completely rigid, as in the **sutures** of the skull. However, RA only occurs in fully movable joints, the so-called **synovial joints**. Each synovial joint consists of two **bones**. These are covered with **cartilage**, where they interact with each other. A **synovial capsule** surrounds the complete joint, and **synovial fluid** fills the space between the bones within the capsule. These synovial fluids along with the cartilage coating are excellent lubricants that allow movement of the joint with minimal friction. If you warm up before any sport exercise, you also warm up the joints. When the synovial fluid gets warm, it becomes thinner and is more easily absorbed by the cartilage. These, in turn, swell and become more flexible, which protects your cartilage from too much wear and tear during exercise. Unlike bone tissue, the cartilage of a joint does not have any blood vessels. It gets its nutrients from the synovial fluid. For a healthy joint, regular exercise is necessary because the movements put pressure on the cartilage and that helps exchange the old synovial fluid within the cartilage with fresh fluid.

What happens to the joint when you have rheumatoid arthritis? The first step in RA is a **chronic inflammation** of the synovial membrane. If it is due to bacteria or viruses in the synovial membrane, this would be temporary. But RA is an **autoimmune disease**, meaning some **antibodies** that are produced by the immune system do not distinguish correctly between the body's own tissues and foreign invaders. In RA, the body makes antibodies that attack cartilage in joints. The antibodies then recruit other cells involved in the immune system, which results in inflammation. It is still unknown why this happens. Typically, the synovial membranes of several joints are attacked at the same time. The inflammation causes pain and stiffness, mainly in the morning. The condition improves during the day after using the joint for some time; the warmed up joint moves more easily. But every morning, the pain is back. Sadly, over time, the uncontrolled inflammation progresses from the synovial membrane to the cartilage layers that protect the bones and keep the joint smooth. Once the cartilage becomes inflamed, it is no longer completely smooth. Moving the joint now causes little pieces of cartilage to be rubbed off. Damage to the cartilage takes a long time to heal, and healing is possible only when the inflammation is stopped. If not treated, the cartilage layer is rubbed off completely and bone damage starts. Such damage to the bone is incredibly painful and cannot be reversed. Most damage happens within 2 to 3 years, so treatment has to start early. If left untreated, the affected joints ultimately become disfigured, the bones do not line up naturally anymore, and the joint becomes completely unmovable.

Today it is possible to treat RA effectively if the treatment is started very early in the progression of the disease. One group of medications act like **steroid hormones**: they suppress the immune response so that the immune system does not attack the joints anymore. The disadvantage of this type of treatment is obvious: over time, the beneficial action of the immune system is also suppressed and in the long run, the body cannot fight off viral and bacterial infections efficiently. The goal with steroid treatments is to find a low dosage that prevents cartilage damage without suppressing the immune system's ability to fight infection. A second group of medications influences the rheumatoid immune response much more specifically, meaning the drugs do not have the same side effects as steroids. Unfortunately, the mechanisms of action of these drugs are poorly understood. Several drugs were discovered while treating diseases like cancer or malaria in people that suffered from RA at the same time. Improvement of RA due to treatment is seen only after months. For immediate relief from the pain, **analgesics** (pain killers) are available, but they do not stop the inflammatory response. Antiinflammatory drugs like Motrin provide only moderate relief.

RA is only one in a number of similar rheumatic diseases. The risk for any form of arthritis rises with age. **Osteoarthritis** (OA), also known as degenerative joint disease, is the most common form of arthritis. It affects more than 12% of the population, or 20 million people, in the United States. While OA is not caused by autoimmunity, it still results from inflammation of the joint cartilage. The incidence of OA is related to age, and there is a strong genetic link as well. Ultimately, OA results from inflammation just as RA does; therefore, it is often treated with steroids and analgesics, just as is RA.

Besides heart diseases, arthritis is the next greatest cause of disability. RA can happen in young people, but in this age group, catastrophic joint injuries in sports or accidents are more common. Synovial joints are designed to provide smooth movement, but also contain ligaments to restrict inappropriate movement. The knee, for example, is a **hinge joint** that is designed to open and close like a hinge in a single direction. The two bones that make the knee joint are the **femur** in the upper leg and the **tibia** in the lower leg. These two bones are connected by a series of tough fibrous ligaments that hold the bones in the knee together to prevent the knee from bending side to side or backwards. Any strong blow against a joint can lead to a joint dislocation, what may lead to rupturing of the synovial membrane, rupture of ligaments that stabilize the joint, and damage to the cartilage in the joint. Sustaining a strong impact to the side of the knee joint forcing it inward can result in a tear of the **medial collateral ligament** (MCL). This type of injury is common in contact sports like soccer and football. Historically, such an injury was treated with major surgery designed to repair the torn ligament. More recently, this type of surgery has been done with a series of large needles, which is called **arthroscopic surgery**. Currently, some physicians are advising some people to strengthen their legs and possibly avoid surgery altogether. Contrary to bones, ligaments and cartilage do not have blood vessels inside. Therefore, they always heal slowly. A broken bone needs at least 6 weeks to heal, but a ruptured meniscus or ligament in the knee typically needs 6 months.

Heather's RA was caught early enough that treatment prevented severe joint damage that would have left her completely unable to walk. She remains active, but not without significant pain. There are no longer days when she declines the use of her handicapped parking permit.

Questions about the case:

1. It was mentioned in the case that the human knee has four different cartilages, yet only one was named (the MCL). What are the others, and what do they do?
2. The case mentions that the knee is a hinge joint. What are two other types of movable joints? How do these joints work?
3. In talking about joints, this case completely left out the concept of tendons. What are tendons? What is tendonitis, and how is it treated?

Questions to go deeper:

1. RA is an autoimmune disease. What are some other examples of autoimmune diseases?
2. Women seem to sustain a larger number of knee injuries than men. Is this really true?
3. Besides the suppression of the beneficial immune response, what are some side effects of steroid treatments?

How Europe's Royal Children Died

A Case Study of Hemophilia and Genetics

Prince Leopold was dying. He was not dying of a war wound from an enemy's attack, but from a slight fall that he took while walking. He had led a very sheltered life. He could not play as other children did because whenever he got a minor cut or bruise, it would bleed profusely and it would not stop. Leopold had to be constantly monitored and cared for because one cut or bump could kill him. Even this constant care could not save him; this time nothing could be done. On March 28, 1884, at the age of 31, Leopold, the duke of Albany, son of Queen Victoria of England, was dead. Leopold was not alone. His grandson Rupert, his nephews Leopold and Maurice, and countless other males in his family tree developed this bleeding disease. In fact, through **inbreeding**, the royal families of England, Spain, Russia, and Germany were all affected by this disease. What caused this tragedy, and why was it only found in males?

It turns out that **hemophilia** was the cause of this tragedy. Hemophilia is a **recessive** genetic disorder in which affected individuals are unable to form blood clots. Therefore, if hemophiliacs get even a slight cut or bruise, they risk bleeding to death. In most cases, hemophiliacs don't live past the age of 10. They are unable to clot because they cannot produce a functional protein that is necessary for a blood clot to form. Hemophiliacs can't make this protein because they have inherited two mutated copies of this gene. Hemophilia affects roughly 1:4,000 male children worldwide. The probability of a female inheriting hemophilia is roughly 1:100,000,000. But, the disease is sex linked, which is why the incidence of hemophilia is much higher in males than it is in females. In order to describe how this occurs, we must discuss the genetic mechanism behind sex determination.

In humans, sex is determined by the complement of **sex chromosomes** that a person possesses. Humans have 23 pair of chromosomes, and the sex chromosomes are numbered as the 23rd pair. If a person inherits two **X chromosomes** (one from her mother and one from her father), then she will develop to be female. If a person inherits one X chromosome and one **Y chromosome**, then he will develop to be male. This means that individuals with an XX **karyotype** are female, and individuals with an XY karyotype are male. Since the Y chromosome carries the **genes** that confer male **traits**, only individuals that inherit a Y chromosome can be male. Therefore, when it comes to having a child, there is a 50% chance that it will be male and a 50% chance that it will be female. You are probably wondering how to calculate that probability. Geneticists use a tool called a **Punnett square** (Figure 1). The letters across the top (in the grey boxes) represent the father's two sex chromosomes (X and Y). The letters in the grey boxes across the side represent the mother's two sex chromosomes (X and X). When sperm and eggs are made, the two sex chromosomes separate through a process called **meiosis**, and are deposited into individual sperm and eggs. Therefore, in the case of males, half of the sperm produced will receive an X chromosome and half of them will receive a Y chromosome. In the case of females, they can only have X chromosomes; therefore, every egg must have an X chromosome. The white squares in the Punnett square show the possible outcomes when a sperm and egg unite. As you can see, half of the possible outcomes are male and half of the possible outcomes are female. Since each child is a separate event, the probability of having a girl would stay the same for the second child, even if the first child were a boy (or a girl for that matter).

Now, back to hemophilia. You are probably thinking that the reason that hemophilia is sex linked is that the hemophilia gene is found on the Y chromosome. Since only males have the Y chromosome and only males get hemophilia, this makes perfect sense, right? This is what some early geneticists thought as well. But, upon studying the inheritance of the disease, they found this was not the case.

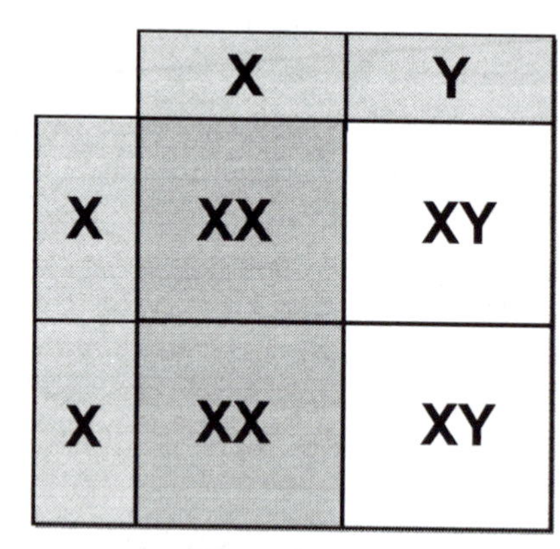

Figure 1

In order to determine the inheritance patterns of certain diseases, geneticists need to follow the disease as it is passed on through generations. The tool they use to accomplish this task is called a **pedigree**. A pedigree is basically a genetic family tree. In this tree, detailed records of a disease within a family are kept, so that geneticists can begin to determine the cause of a particular disease. Figure 2 show the pedigree for hemophilia in the European royal families. In pedigrees, males are depicted as a square and females are depicted as a circle. Individuals with hemophilia are red and normal individuals are white. As you look down this pedigree, you notice something interesting; not every male has the disease. Also, if a male has the disease (e.g., Leopold, son of Albert and Victoria), his sons do not. If the hemophilia gene was found on the Y chromosome, then every affected father would have an affected son, and every affected son would have an affected father. Since a male child can receive his Y chromosome only from his father, he must inherit hemophilia as well. This demonstration shows that hemophilia cannot be transmitted on the Y chromosome. How else could it be sex linked then?

The other option for the location of the hemophilia gene is the X chromosome. But wait a minute; females have only X chromosomes. Why would there be such a low occurrence of hemophilic females? Let's look at another Punnett square to answer this question. The Punnett square in Figure 3 shows the possible outcomes of breeding between Queen Victoria and Albert (Leopold's parents). Assume Queen Victoria was a **carrier** of hemophilia. This means that she would have 1 X chromosome that is normal and produces a functional clotting protein (X), and 1 X chromosome that is mutated and carries a mutated hemophilia gene (X*). Since she has one copy of the normal clotting gene, she produces enough of the functional protein, so she has no signs of hemophilia. But if you look at the outcome of this cross, you notice something interesting. Half of the males receive the mutated hemophilia gene (X*). Since

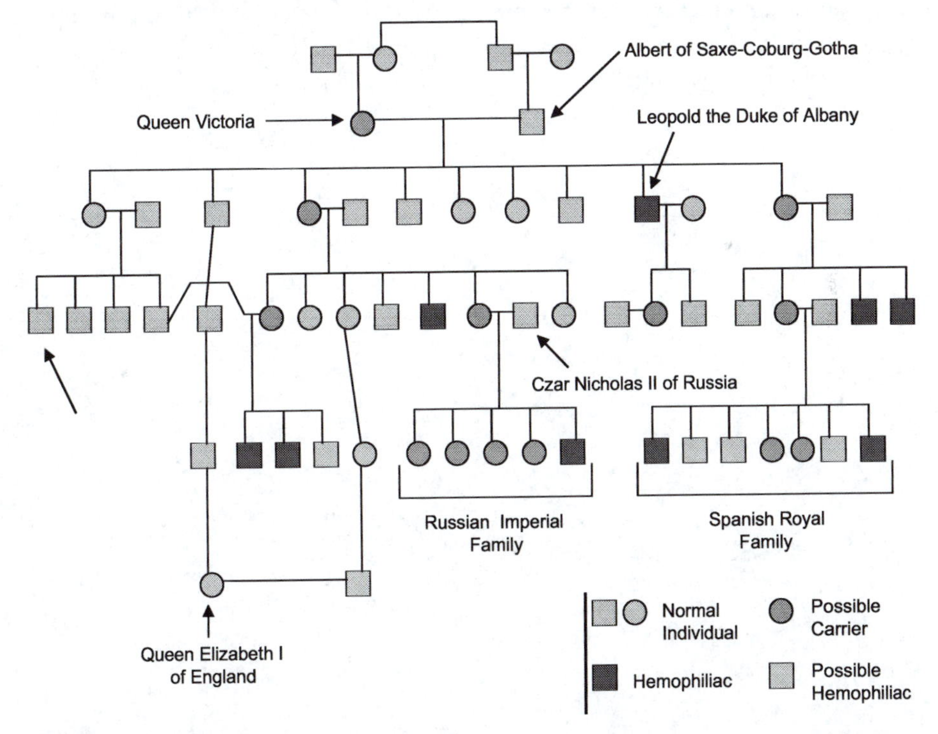

Figure 2 This pedigree shows the inheritance of hemophilia in the European royal families. Males are represented with squares and females are represented with circles.

we are assuming that the X chromosome is the location of the hemophilia gene, this means that these males (X*Y) only have one copy of the clotting protein (the faulty one). Since these males only produce faulty clotting protein, they cannot clot. Thus, they have hemophilia. From looking at the Punnett square you can also notice that one half of the females also receive the mutated clotting gene (X*X). This female would also be a carrier of the disease because she has one bad copy and one good copy of the clotting protein. Her sons would then have a 50% chance of having hemophilia, but her daughters could only be carriers. As you look through the pedigree for hemophilia in the royal families, this mode of inheritance makes sense and allows for an accurate explanation for the sex-linked inheritance of hemophilia.

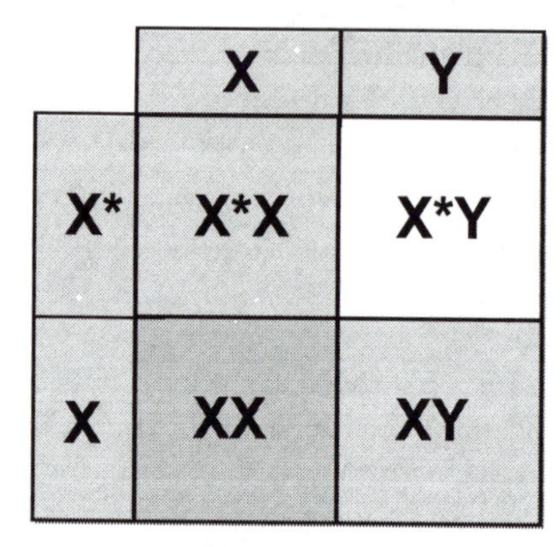

Figure 3

As sad as this situation was, the high incidence of hemophilia in the European royal families can be completely attributed to inbreeding. If the royal families had not tried to keep the "royal lines" pure, many sad deaths could have been avoided.

Questions about this case:

1. Why did Queen Victoria have only one affected son with hemophilia? (Wouldn't 50% of her sons have hemophilia based upon the Punnett square?)
2. How would it be possible for a female to develop hemophilia?
3. How did Queen Victoria become a carrier of hemophilia?
4. How would the inheritance of hemophilia be different if it were Y-linked dominant? All males would get it

Questions to go deeper:

1. What are some other X-linked traits? color-blindness
2. How is hemophilia treated today? helping with clotting
3. Are there Y-linked traits? If so, give an example.

fewer active genes & less likely to be mutated

References:

http://www.nlm.nih.gov/medlineplus/hemophilia.html

http://www.sciencecases.org/hemo/hemo.asp

Snustad DP an Simmons MJ (2006) Principles of Genetics 4th ed. John Wiley and Sons, Inc. Hoboken, NJ.

A Tribe Divided

A Case Study of Diabetes and the Pima Indians

Few people know about the Pima Indians. For thousands of years, the people of this Indian Nation lived along the Gila River between Phoenix and Tucson in Arizona. They have a legacy of being great engineers, master farmers, and experts in weaving and basketry. Their elaborate system of water channels made the desert bloom and allowed them to grow even water-intensive crops in the desert. They also have been a remarkably friendly tribe, showing great hospitality to any traveler. Today the Pima people are still mainly farmers, though their crops have changed from corn and wheat to oranges and olives. We study the Pima Nation because they are outstanding in a quite disturbing way: more than 60% of all Pima between 20 and 80 years of age have diabetes. Compare that to the **prevalence** of diabetes in the United States at about 8%, and worldwide at about 5%.

What could be the reasons for such an extremely high rate of diabetes in this Indian Nation? For centuries, the people of the Pima married other Pima. Therefore, the people of the Pima are genetically very similar. Diabetes can be traced in many of their families over several generations. It could well be that they possess **genes** that favor the development of diabetes. Yes, the Pima are masters of agriculture, but still these people lived in a harsh environment for thousands of years. Genes that help them to store energy in their body or that might cause them to burn energy more slowly would have been a tremendous advantage to survive in such an environment. Today, however, the same genes might be related to diabetes. Scientists have confirmed that diabetes can, indeed, have a genetic link in some people. Besides genes, scientists have also shown that diabetes is linked to obesity. This means that a diet rich in high-fat and high-calorie foods together with a sedentary lifestyle can lead to obesity and an increased risk of contracting diabetes. Before we discuss what has been found regarding why the Pima Indians suffer so often with diabetes, let's discuss a few basics of diabetes that will help us understand this research better.

There is always a certain level of the sugar **glucose** in our bloodstream. It is essential for all of us. Our body, especially our brain, needs glucose as a source of energy. Our body keeps the amount of glucose in the blood constant and at an optimal level using **negative feedback loops**. If you eat something sweet, for example frosted donuts, your intestines take up all the glucose molecules contained in this meal and deliver them into the bloodstream. In a short time, your blood is swamped with glucose. The glucose level is constantly measured by cells in your **pancreas**. As soon as the **blood glucose level** rises, the cells in the pancreas secrete the **hormone insulin** into the bloodstream. Insulin itself does not do anything to the glucose in the blood, but this hormone is understood by all healthy cells in the body as the signal to act. When insulin binds to receptors on cells, especially those of the liver, the cells start taking up glucose from the bloodstream and, in a short while, all the surplus glucose is removed from the bloodstream. As the blood glucose level decreases, the pancreas stops releasing insulin. Consequently, the cells soon stop taking up glucose.

In a person with diabetes, two things may have gone wrong. First, the pancreas may have become unable to produce insulin. If that is the case, and a sugar-frosted donut is eaten, the blood sugar level rises, but there is no insulin to tell all the cells in the body to take up the surplus glucose. The person becomes **hyperglycemic**; that means the glucose level in the blood is elevated for long periods of time. Since the pancreas does not produce enough insulin, this is called "**insulin-dependent**" or "**Type 1**" **diabetes**. Most often, there is a genetic reason for this defect in the pancreas, and onset of the disease is early in life, often as a teen. This type can be managed, but not permanently cured, by injecting adequate amounts of insulin into the bloodstream after each meal. In this situation, the diabetic will routinely test his or her blood glucose levels with a pocket-sized device so that the proper amount of insulin can be injected. There is a second way that diabetes develops, and in this case, the pancreas produces normal amounts of insulin. As soon as the glucose level rises, the pancreas releases insulin as it should. However, in this type of diabetes, most of the body cells do not have adequate receptors for insulin. Healthy cells have a receptor protein in their cell membrane that can respond to insulin. If the receptor protein encounters insulin, then it starts all the processes necessary to take up glucose inside the cell. In "**Type 2**"**diabetes,** the receptor proteins on the surface of cells either are not being produced in high enough quantities or have become defective. Insulin arrives at the cells, but the receptors for insulin do not function properly, and consequently, there is little or no response. The cells do not take up glucose at a normal rate, and the blood remains hyperglycemic. Type 2 diabetes usually has no clear genetic cause, but is mainly the consequence of poor dietary choices that lead to obesity. Type 2 diabetes is not easily treatable, and disease management requires dramatic changes in lifestyle and eating habits. Unlike Type 1 diabetes, Type 2 can be reversed in some cases if the dietary and obesity issue is corrected.

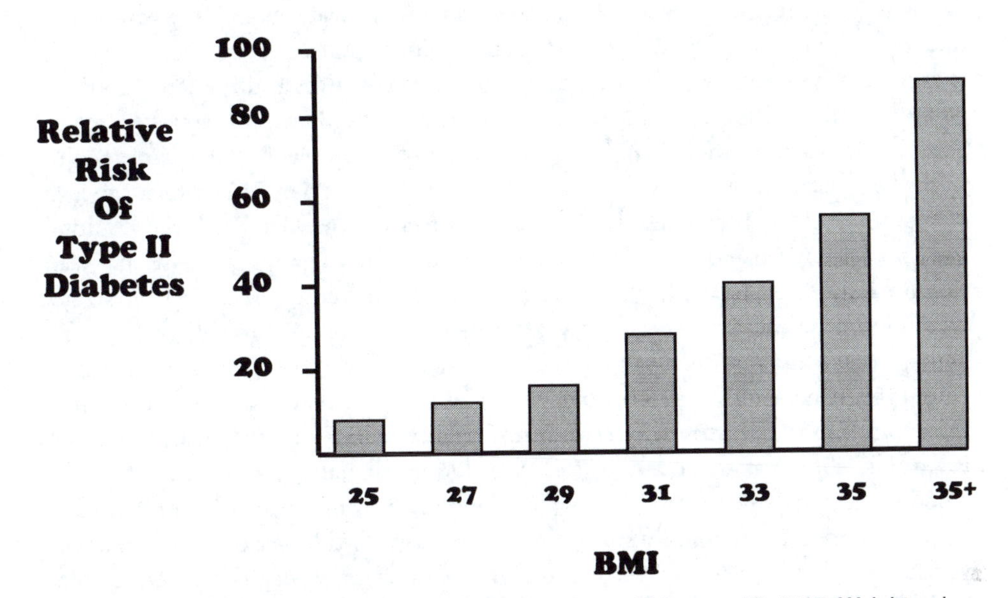

Adapted from Colditz, GA., Willett, WC., Rotnitzky A., and Manson JE. 1995. Weight gain as a risk factor for diabetes mellitus in women. Annals of Internal Medicine. 122:481-486

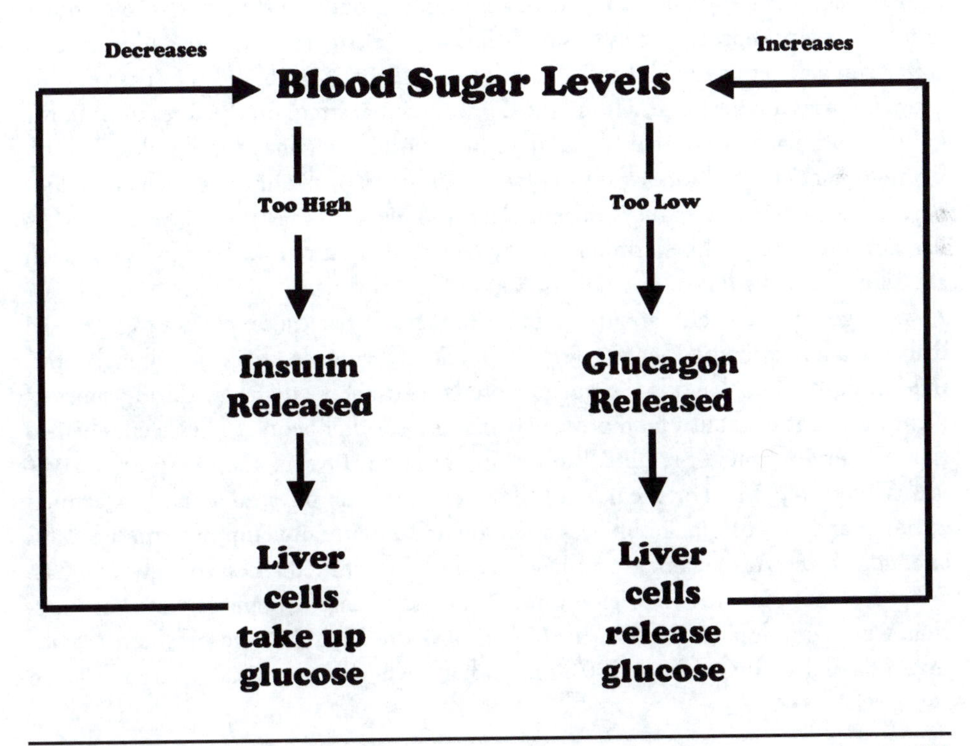

As you can see, both types of diabetes lead to **hyperglycemia**. But why is too much glucose in the blood dangerous? After all, don't we need glucose as a source of energy anyway? People with hyperglycemia are unusually thirsty, drink large amounts of water, and have the need for frequent urination. They can get severe headaches, have difficulty concentrating, and frequently experience problems with their eyesight. In the long term, the tiny blood vessels in their eyes become damaged, which can lead to blindness. If the blood vessels in a diabetic's feet become damaged, the resulting poor circulation can allow even minor infections to result in **gangrene.** This may lead to limb amputations. People with hyperglycemia often feel weak and tired, they get skin infections, and any wound they get may seem to refuse to heal. Even worse, hyperglycemia usually does not come alone. Ninety-five percent of all people who have Type 2 diabetes are also **overweight** or **obese,** and they very often also suffer from **high blood pressure (hypertension)**. Diabetes with obesity and hypertension is a deadly combination because it can cause kidney damage. The **kidney** filters the blood to clean it of waste products, which the body cells produce. The **capillaries** in the kidney are microscopically small filter units. How do they get damaged? Imagine you use a paper filter for coffee. It is designed so that as water drips on it, it holds back the pieces of ground coffee beans, but it lets through water and the tasty molecules that make up the coffee drink. Imagine you take such a paper coffee filter and put it in a machine that not only drips water on it, but pushes water through with high pressure. Additionally, you do not use only ground coffee, but whole coffee, beans. In a very short time this misuse will cause the paper to rupture and instead of coffee you will get an undrinkable mess. If someone has high blood pressure and diabetes, the risks are the same. The elevated pressure puts strain on the filter units in the kidneys and diabetes bombards the tiny filters with large glucose molecules. The filter membranes in the kidney rupture. Even if only 5% of the filtering cells in the kidneys are damaged, the kidneys do not clean the blood efficiently anymore. At 10% damage, the result will be terminal kidney failure. At that point, the only choices left are **dialysis** for the rest of one's life or a kidney transplant.

Let's go back and talk about the Pima Indians. What causes such a high rate of diabetes among them? Genetic testing provided evidence that the Pima people, indeed, have genes that allow them to store fat extremely efficiently during times of plenty and that they burn fat more slowly than most other people. These genes helped them to survive times of famine. But during the last 150 years, they adopted the typical Western lifestyle. For them, work now requires little physical activity. Machines at home and at work make life less strenuous. Cars virtually eliminate the need for strenuous hikes. At the same time, they have nearly unlimited access to high-fat foods. What has more impact: their environment or their genes? To answer this last question, we need to look at Mexico. In the Sierra Madre Mountains in northern Mexico lives a second group of about 700 Pima Indians. They are genetically very similar to the American Pima, but they still live a much more traditional lifestyle. They eat more low-fat foods and while American Pima spend on average 3 to 12 hours per week in moderate or strenuous physical labor, the Mexican Pima spend 22 to 33 hours per week in such types of work. In spite of having the same genes as the American Pima, the prevalence of diabetes in the Mexican Pima is only around 6%. While more than 90% of all Pima in the United States are considered overweight or obese, only 30% of the men and 60% of the females among the Mexican Pima fall in the same groups.

In conclusion, we can say that even if a person is genetically prone to develop Type 2 diabetes, this disease primarily develops when the environment and the person's behavior allow it. It seems quite obvious that most cases of Type 2 diabetes are completely preventable by changes in lifestyle and selection of high-fiber and low-calorie foods. Can you imagine that your chances of one day spending hours in a dialysis center starts today with your decision of whether to walk instead of drive?

Questions about the case:

1. The case describes the regulation of blood sugar levels as a regulatory feedback loop. What does this mean?
2. Your pancreas should release insulin in response to rising blood sugar. What hormone does your body release in response to dropping blood sugar?
3. If your body is unable to manage dropping blood sugar levels, what disease do you probably have? What symptoms would you have, and how would this be treated?

Questions to go deeper:

1. What is homeostasis, and how does it relate to diabetes?
2. Besides blood sugar levels, what other factors are kept constant in our bodies?
3. What are the recommended amounts of food and physical activity to avoid obesity and diabetes?

References:

1. Schulz, L. O., Bennett, P. H., Ravussin, E., Kidd, J. R., Kidd, K. K., Esparza, J., and Valencia, M. E. (2006) Effects of traditional and western environments on prevalence of type 2 diabetes in Pima Indians in Mexico and the U.S. Diabetes Care. 29: 1866–1871.

2. http://www.diabetes.org/ The American Diabetes Association

3. http://www.diabetes.niddk.nih.gov/dm/pubs/pimsa/index.htm "The Pima Indians: Pathfinders for Health" National Institute of Diabetes and Digestive and Kidney Diseases

4. http://www.hhmi.org/biointeractive/obesity/pima_clip.mov Howard Hughes Medical Institute, 3-minute film clip on the Pima Indians

Joe's Favorite Drink

A Case Study of Phenylketonuria

Joe was thirsty after a hard day of yard work on a warm June day. Whenever he had a thirst like this, he absolutely had to have his favorite drink: an ice-cold Diet Coke. So, he jumped into his truck, raced over to the local AM-PM, picked up an ice-cold 1-liter bottle, and began to drink it like there was no tomorrow. On his way home, Joe noticed something interesting on the ingredients label that he had never seen before: **Phenylketonurics:** Contains **Phenylalanine**. "What the heck is a phenylketonuric?" Joe wondered. "And what is phenylalanine?"

Well it turns out that the correct question is not "*What* is a phenylketonuric?" but "*Who* is a phenylketonuric?" A phenylketonuric is a person who suffers from the **recessive** genetic disorder **phenylketonuria (PKU)**. PKU is a disease that affects the body's ability to **metabolize** specific **amino acids**. If left untreated, PKU can lead to severe mental retardation and epilepsy. PKU is a relatively rare disease that occurs in 1:12,000 live births worldwide. A person suffers from PKU when he or she receives two **mutated** copies of the **phenylalanine hydroxylase gene** from his or her parents. Every person normally has 46 chromosomes, which are arranged into 23 **homologous** pairs. For each pair, one chromosome is inherited from the mother and one is inherited from the father. In the case of PKU, the affected child receives two copies of chromosome #12 (one from mom and one from dad) that each should contain a copy of the functional phenylalanine hydroxylase gene, but, instead, each of these copies is mutated. Since both copies of the gene are defective, the child is unable to produce functional phenylalanine hydroxylase protein. It is the body's inability to produce normal phenylalanine hydroxylase protein that can lead to the symptoms of PKU.

Phenylalanine hydroxylase is an **enzyme** that is required for the processing of the amino acid phenylalanine. Enzymes are the "chemical motors" that speed up (**catalyze**) chemical reactions that occur in the body. In some cases, enzymes speed up the rate of a reaction by as much as 10 million times its normal rate. Without the action of enzymes, most chemical reactions would occur at a rate that would not support the needs of a living organism. Enzymes speed up the rate of a reaction by providing an alternate pathway that lowers the **activation energy** of the reaction (Figure 5). Since less energy is required to "start" a chemical reaction, it can occur at a much faster rate. It is important to note that an enzyme can only lower the activation energy of a reaction: it cannot make a chemical reaction that is **energetically unfavorable** become favorable.

Enzymes also perform their duties in a very specific manner. In most cases, a specific enzyme can only catalyze a specific chemical reaction. This means that a specific enzyme can only accept and act upon certain chemical compounds or **substrates**. It is believed that this occurs because only certain chemicals can fit into the **active site** of a specific enzyme. Since most chemicals have unique shapes and charges, only those chemicals that are attracted to the shape of the active site can be processed by the enzyme. This idea is called the **lock and key** model of enzyme specificity. In the case of phenylalanine hydroxylase, it catalyzes the chemical reaction in which the amino acid phenylalanine is the substrate (it is the only molecule that fits into the active site) and the amino acid tyrosine is the **product**. So what exactly is an amino acid?

Cells link amino acids end to end in order to form **proteins**. Amino acids are carbon-based molecules that are characterized by three specific groups: an **amino group**, **a carboxyl group**, and a **side chain.** During protein formation, the amino group of one amino acid and the carboxyl group of the next amino acid are modified and linked together by forming a **peptide bond**. The side chain is what gives each one of the 20 amino acids used by the human body its unique characteristic. In addition to being the building blocks of proteins, amino acids can also be metabolized to perform various tasks in the body. In the case of phenylalanine, it is often modified, through a variety of chemical reactions, to become **melanin**. Melanin is the pigment found in skin and hair that gives these structures their color. Phenylalanine metabolism can also give rise to a variety of **neurotransmitters**. Neurotransmitters are the chemicals that transmit neural messages across **synapses**. In each case, the chemical reaction that changes phenylalanine into tyrosine begins the modification process.

Even though amino acids are necessary for cells to function, the human body is only able to produce 12 of the 20 amino acids on its own. These are known as the **nonessential** amino acids. The other eight **essential** amino acids have to be acquired through the foods that we eat. Because phenylalanine is one of the essential amino acids, it is a necessary dietary component for proper nutrition. You are probably now wondering, "If we need to have some phenylalanine in our diets, why do people with PKU need to know if a product contains phenylalanine?"

Remember that people with PKU lack a functional phenylalanine hydroxylase gene. Without this gene, the body cannot produce a functional phenylalanine hydroxylase enzyme. Without this enzyme, the body in unable to carry out the chemical reaction that changes phenylalanine into tyrosine. Therefore, when someone with PKU ingests phenylalanine, the body is unable to metabolize it; so, it just builds up. There are two major consequences to this build-up. First, phenylalanine can undergo a chemical reaction where it is turned into **phenylpyruvic acid**. It turns out that phenylpyruvic acid is highly toxic to the nervous system. Therefore, when a person with PKU ingests large amounts of phenylalanine, the accumulation of phenylpyruvic acid in the nervous system leads to **neuron** death. Second, when the amounts of phenylalanine and tyrosine become out of balance in the brain, the production of neurotransmitters decreases. This means that the communication between neurons in the brain becomes impaired. In both cases, the result is mental retardation and epilepsy. Therefore, the more phenylalanine that a person with PKU ingests, the worse the symptoms become. However, if a low-phenylalanine diet is implemented early in life, the symptoms associated with PKU do not develop. In order to diagnose PKU as early as possible, children in the United States are tested at birth using a special blood test. If a positive diagnosis is reached, then the parents are sent to a dietician so that the child can be placed on a low-phenylalanine diet. Since it can be hard to keep track of every food that is high in phenylalanine, most companies place warnings on their products, including Diet Coke.

Fortunately for Joe, he inherited at least one normal copy of the phenylalanine hydroxylase gene. Therefore, even if he drank 2 liters of Diet Coke every day, he would not develop symptoms of PKU. But, the effect of the amount of caffeine in 2 liters of Diet Coke is a completely different story.

Questions about this case:

1. If a normal person (non-phenylketonuric) does not ingest any phenylalanine, what would be the consequences?

2. If a normal person (non-phenylketonuric) ingests a lot of phenylalanine (i.e., drank 2 liters of Diet Coke), what happens to the excess phenylalanine?

3. Are there any enzymes that have multiple substrates?

Questions to go deeper:

1. There are many other diseases that are caused by mutated genes. What are the names of these diseases? What genes are mutated in these diseases?

2. What is a complete protein? What are some good sources of complete proteins? Why do people on strict vegetarian or vegan diets need to pay attention to the sources of complete proteins?

References:

http://www2.coca-cola.com/contactus/faq/labeling.html

http://wikipedia.org/wiki/Phenylketonuria

http://www.ncbi.nlm.nih.gov/disease/Phenylketo.html

And the Winner is... Dead!

A Case Study of Blood Pressure and Homeostasis

Some people in a small village in India thought of a splendid idea for some real fun: a salt-eating contest. The contest attracted hundreds of villagers. Salt is essential for the human body, but as everybody in the contest soon found out—too much of it can be dangerous for your health. Jiten B., 19, became ill after consuming nearly 1 pound of salt during the contest. Evidently, he passed out at the table—causing the contest to be abandoned altogether. Sadly enough, it did not help Jiten. He was pronounced dead soon after they got him to the hospital. But what was the reason for Jiten's death? How can something that we absolutely need for survival kill us?

To understand what killed Jiten, we need to understand a basic concept of physiology called **homeostasis**. Homeostasis is the tendency of our body to keep the internal environment constant despite changes in the external environment. We all know this fact from our body temperature. Regardless of whether you are in the middle of the desert at noon with a temperature of 60 °C (140 °F) or in an ice-cold night in Alaska at midnight with at a temperature of –60 °C (–76 °F), your body will try to keep your internal temperature very close to 37 °C (98.6 °F). If you are, indeed, in the desert at high temperatures, you will sweat profusely so that evaporation of the fluid can cool you down. If you walk outside at freezing temperatures, your body will first conserve heat by keeping the blood away from your skin, so that the heat is not easily lost into the environment. If that is not sufficient, you will start to shiver, so that the muscle movements can provide extra heat to warm your body.

What about salt and homeostasis? Your body works to keep the concentrations of sodium constant in the **cytoplasm** inside your cells, in the **blood plasma**, and in the **interstitial fluid** between the cells. The amount of sodium in blood plasma and around the cells typically is 30 times higher than inside the cells. This concentration difference across the cell membrane is necessary for all cells to function normally. Specifically, our neurons and muscle cells would be affected negatively if the difference gets too high or levels out. Each cell works continuously to keep up this gradient. In fact, about 50% of all calories you eat are used to drive the pumps that keep sodium out of the cell. Additionally, when you consume sodium with your normal diet, the **intestines** and **kidneys** respond in ways that keep the concentrations in the body constant. During the course of a normal day, the intestines absorb sodium from the food, and the kidneys excrete an equal amount of sodium into the **urine**. If you eat very little sodium with your diet, then the intestines take up sodium more efficiently, and at the same time, the kidneys reduce the loss of sodium into the urine. This results in a constant concentration of sodium in our blood plasma and around our cells.

Usually the concentration of water in our cells (amongst all the other molecules and ions that are dissolved in it) is exactly the same as the concentration outside the cells. This condition where the salt concentration outside the cell is equal to the concentration inside the cell is called **isotonic**. Eating nearly 1 pound of salt in the course of a salt-eating contest overwhelms all regulatory mechanisms for sodium. The intestines try to keep sodium out of the bloodstream, but there will be so much around, they cannot hinder sodium ions from flooding the blood plasma. The kidneys immediately increase the loss of sodium into the urine but this mechanism works too slowly. Very soon, the blood plasma and the fluid around our cells contain much more than the normal concentration of sodium. The fluid bathing the cells now is **hypertonic**; outside the cells there are many more sodium ions dissolved per unit of water than inside the cells. Most body cells have no mechanism to hinder water crossing the cell membrane. Such a hypertonic fluid around the cells literally sucks water out of the cells, and the cells begin to shrink because they lose water. Sodium ions will leak into the cells more quickly than the cells can pump them out again. Shrinkage and increased sodium in the cells can quickly inhibit normal cell function and irreversibly damage human cells. The neurons in our brain are most sensitive to any such changes; very soon the person becomes confused and may fall into a coma. If the centers of the brain that coordinate the muscles needed for breathing or regulate the normal heart rate stop working, then the person stops breathing or gets a completely irregular heart beat. Each of these possibilities can result in death.

For you and me, the major concern is not about eating too much salt during a contest. Instead, for us, the greatest danger related to salt comes from eating too much salt with our normal diet. In a healthy body, the normal regulatory mechanisms keep the salt and water concentrations balanced over a long period of time and during various amounts of sodium intake. The body uses two separate mechanisms to regulate sodium and water, but they work closely together to control blood pressure. A constant, excessive, sodium intake can be handled by the body by increasing the rate of sodium excretion by the kidneys and/or by retaining more water in the body. If the elevated sodium intake is more than the kidneys can excrete, then the only way to keep the concentrations equal is to retain more water. The additional water stays mainly in the circulatory system and increases the blood volume. The increase in blood volume makes the heart pump more blood through the body. Additionally, salt causes the blood vessels to constrict in sodium-sensitive people, which leads to an increase of blood pressure, called **hypertension**. Because hypertension develops almost exclusively in communities that have a fairly high intake of salt, the average exceeding 6 grams daily, scientists conclude that salt intake is an important risk factor for the development of this disease. This link has been confirmed in numerous scientific studies (1). To be complete, you need to know that other risk factors for hypertension exist. For example, genetics plays a role, as well as age, and there is no way to change these things. Obesity is one more highly important risk factor and it can be controlled. If you add just one pound of body weight in fat, then your body has to add several miles of blood vessels (mainly capillaries) just to keep the new fat tissue alive. This automatically means that your heart has to pump harder to get the blood through the added vessels.

Hypertension is defined as sustained, elevated **blood pressure** in the arteries of 140/90 mmHg, or higher. Normal blood pressure is 120/80 mmHg, or lower. The first number indicates the pressure when the **ventricles** of the heart contract and the blood is pushed out of the heart into the arteries. This is called the **systolic pressure**. During the time needed to fill the ventricles again with blood, the arteries contract back to their starting diameter, pushing the blood further on. This is the **diastolic pressure**, given as the second figure of the measurement. The expanding and contracting of blood vessels is what you can feel as the pulse. Together, these numbers indicate how hard the heart must work to keep the blood circulating through the body. In adults, the heart beats 80,000 to 90,000 times a day, and a relatively small increase in strain can have severe consequences. Hypertension affects about 75 million Americans, that is about 1 in 3! Every person with hypertension lives with an increased risk for **coronary heart disease**, heart failure, **strokes**, and kidney failure. Approximately 1.1 million deaths in the United States were related to these diseases in 2003. The average sodium intake in the United States is between 5 and 6 grams per day. If cultures with high sodium intakes, like the U.S., could reduce that to 2–3 grams per day, then half of the people who are on hypertension medications today could stop taking their pills, and the number of deaths caused by strokes and heart disease would each be reduced by 20%. In other words, 200,000 people would still be alive if they had taken a low-salt diet seriously (2). As you can see, too high a dietary salt intake can be nearly as dangerous as taking part in a salt-eating contest.

If you think an accident like that in the salt-eating contest could only happen in a country like India, then you'd better think again! In January of 2007, KDND 107.9 radio station in Sacramento, California, held a water-drinking contest where the contestant who drank the most water without urinating would win a new Nintendo Wii video game system. A few hours after the contest, it was discovered that one of the contestants had died. The suspected killer was water intoxication. Most people (including the contest organizers) had no idea that drinking too much water could kill. This should remind us that an extreme amount of even a good thing can be toxic.

Questions about this case:

1. What other minerals, besides sodium, does our body needs?
2. What is the recommended daily salt intake for a healthy diet?
3. Name 3 food items that are high in sodium.
4. What are some ways you can reduce your sodium intake?

Questions to go deeper:

1. What is an isotonic sport drink? Why is such a drink ideal for balancing water and salt loss during high body activity?
2. What are the main parts of the kidney, and what are their main functions?
3. People who have serious kidney disease may lose function of both kidneys and end up on "dialysis." What is dialysis?
4. What are kidney stones? How do they develop, and how are they cured?

References:

http://www.who.int/cardiovascular_diseases/guidelines/hypertension_guidelines.pdf (free pdf from the article published in Journal of Hypertension, Vol. 21, pp 1983–1992)

WHO/IHS hypertension guidelines

http://www.americanheart.org/downloadable/heart/1136308648540Statupdate2006.pdf

1. Tuomilehto, J., et al. (2001). Urinary sodium excretion and cardiovascular mortality in Finland: a prospective study. Lancet 357:848–851.

What is the Name of Your Husband? "Mary, I Think."

A Case Study of Alzheimer's Disease

This was the most difficult part to deal with for my grandfather: his beloved wife for 48 years did not remember his name anymore. Only 3 months ago, he noticed that she was having difficulties balancing a check book—something she had done without a flaw since they were married. Then she occasionally forgot to prepare dinner. As time went by, she forgot more simple tasks. It really began to worry her, even admitting once that she thought she was going crazy. Together they went to a doctor, and he immediately suspected **Alzheimer's disease**. They scheduled an appointment with a neurologist 2 weeks later to confirm the diagnosis. The day of the visit to the specialist, Mary greeted her husband at home with, "Who are you? How did you get in here?" A few minutes later, she remembered him again—but he was still shaken when they arrived at the hospital. The doctor asked, "What is your name?" Her reply was, "Mary." The doctor then asked, "What is your last name?" Her reply again was, "Mary." Next the doctor asked, "What is the name of your husband?" Again her reply was, "Mary, I think." The neurologist calmly asked a second time, "What is your husband's name?" "Oh, my husband…", but at that point, she could not remember. Just 2 months later, my grandmother, Mary, could not even remember her own name. All the medications she took seemed to have no effect. Another 4 months later, she was not able to communicate with anybody; she became more and more bedridden and needed more and more help for daily activities like dressing and feeding. She finally died of Alzheimer's disease 3 years after the initial visit at the hospital.

What happens during Alzheimer's disease that first causes memory loss and finally results in the loss of the abilities that are necessary for an independent life? Alzheimer's is a **neurodegenerative** disease, meaning that nerve cells in the brain slowly become damaged and later die. **Neurons** in the brain have no ability to divide, and there are no adult **stem cells** in the brain that could replace them. If neurons die, the function, skill, or memory the neurons support is lost. All important functions in our brain have backups, so loss of a few cells can easily be compensated for. However, in progressive Alzheimer's disease, the number of cells in the brain gets so low that effects become noticeable.

In the healthy brain, each neuron receives input either from sensory organs or other neurons. The **dendrites** are the specialized processes of the nerve cell to receive such information. The cell body or **soma** integrates all the information that can lead to a specific response of the neuron in the form of a series of **action potentials**. An action potential is a wave of electricity that travels down the **axon** to the **synapse**. When an action potential reaches a synapse, a cascade of cellular events starts. That ultimately leads to the release of **neurotransmitters** into the synaptic cleft. The dendrites of the next neuron in line have receptors for these neurotransmitters. The soma integrates the information transmitted by this and other synapses that can lead to the next response. In Alzheimer's disease, an **enzyme** in the cells starts to malfunction: a small protein is produced that has no function for the cells and that cannot be easily removed. This protein accumulates both inside and outside the cells; **plaques** appear in areas of the brain with no remaining living cells. **Tangles**, the protein accumulations inside the neurons, become visible in microscopic images of the brain. Both are typical signs of Alzheimer's, but are only detectable in postmortem examinations. The more tangles present in a cell, the more the normal function of the cell becomes disabled. The cells typically respond as if they have **inflammation**. The inflammatory-like response then also affects neighboring neurons and **glia cells**, so that the **homeostasis** of whole brain areas is disrupted. Not all neurons are equally affected. This explains why in some brain areas, like those for memory, the damage appears first. Later, the brain areas for speech and body coordination become affected. In a very late stage, the brain areas that coordinate body functions like breathing and heart rate are destroyed, and the person dies. In the end stage, each person with Alzheimer's becomes completely helpless and needs total care. Depending on the person, the illness may lead to death within 3 to 20 years after the onset of the first symptoms.

A similar neurodegenerative disease is **Parkinson's**. In the healthy brain, some neurons produce the neurotransmitter **dopamine**. Like all neurotransmitters, dopamine travels across the synapse and binds to receptors on neighboring dendrites. But, instead of causing the neighboring neuron to increase the number of action potentials, dopamine causes a decrease in the activity of the target cells. The area of our brain responsible for controlling movements needs certain amounts of dopamine to function normally. As long as the dopamine level is high enough, the cells control each other and work together to produce a smooth movement of muscles. However, in patients with Parkinson's disease, less than a quarter of the dopamine-producing cells still work. All the others are damaged, degenerated, or dead. This causes the nerve cells that control the movement to become overactive. This over-stimulation makes it progressively more difficult to control movements of the muscles those cells are responsible for. Symptoms include the common trembling of hands, arms, legs, and the face, or the limbs that feel stiff and movement that becomes slow. Finally, Parkinson's patients often have problems with balance and body coordination.

There is no cure for Alzheimer's or Parkinson's disease. Medications used to treat most neurological diseases typically increase the amounts or lifespan of neurotransmitters that are released by the cells to help retain proper function of the neurons. This increase in neurotransmitter concentration helps with retaining memory and allows the patients to stay independent for several months or years. But, these medications cannot provide neurotransmitters when all the neurons are dead in Parkinson's disease or slow or break down the production of protein tangles in Alzheimer's disease. Tragically, as both diseases progress, they debilitate the patient gradually and ultimately cause the death of the afflicted person.

Both diseases affect men and women in similar numbers and develop mainly in people over 65 years old. About 4.5 million Americans live with Alzheimer's disease today. Back in 1980, that number was half as high. Interestingly, the disease has a higher prevalence in technologically developed countries than in most countries in Africa and less-developed countries of Asia (1). Parkinson's disease shows no differences between social or ethnic groups. Approximately 1.5 million Americans live with Parkinson's disease. The forecast for both neurodegenerative diseases is grim: by 2050, the number of Alzheimer's cases will have risen above 10 million people, and more than 5 million are expected to suffer from Parkinson's disease. This is mainly due to rising life expectancy. If any drug could be found that delays the onset of the diseases by a mere 5 years, then the number of cases would sink approximately 50% (2).

Questions about this case:

1. What are the major parts of a neuron and their functions?
2. What are the major parts of the brain and their functions?
3. There are some drugs to help those who have Parkinson's disease. What do these drugs do?

Questions to go deeper:

1. Today there is no cure for Alzheimer's disease; the best medications can relieve the patient of some symptoms for a while. But, scientists are close to developing tests for early detection of the disease. Would you want to get tested?
2. What are the events during an action potential and at the synapse that lead to transmission of information?
3. What are some other common neurodegenerative diseases and their symptoms?

References:

http://www.alz.org/

http://www.parkinson.org/

Sex in a Bottle

A Case Study of the Sense of Smell

> A new sexual revolution: men's cologne containing genuine female sex pheromones! Do you want to be more popular with women? You think that is impossible? Here is our guarantee: either you are amazed by our cologne and how it changes your life, or you will get your money back — no questions asked!
>
> Pheromone expert Dr. Tado explains: Males produce a sex pheromone that attracts the opposite sex. We succeeded in recreating the natural pheromone in our laboratory. In nature it is only found in minute amounts in the sweat of men. Way back in evolutionary times, males depended on these pheromones to attract mates. Today you can buy the pure pheromone in much higher concentrations—and the good news is, it still works!

Are these claims true? If they were, would those of you who are men buy the cologne? If these claims were true, how would you women feel about being attracted to a man simply because of the cologne he was wearing? To evaluate the claims of this advertisement, we have to investigate how our chemical senses work. Then we have to review reliable research articles that report results from experiments on human pheromones.

In general, each **sensory organ** receives stimuli from the environment and translates these stimuli into a series of electrical impulses that are sent to the brain for analysis. This translation happens in the sensory cells in our sense organs. Scientists call the events involved in the translation of a stimulus into electrical impulses the **sensory transduction cascade**. The endpoint of the transduction cascade always is the opening of **ion channels** in the **cell membrane**. Ions crossing the cell membrane mean that electrical charges are transported across the membrane. Thus the elements of the transduction cascade together to translate a specific external stimulus into an internal electrical signal.

Each of our senses has one **adequate stimulus**. This is the stimulus that most easily activates the sensory transduction cascade. For example, the sensory cells that are most easily activated by electromagnetic waves called light are found in the retina of the eye. However, sensory cells can also be activated by **inadequate stimuli**. If someone punches you in the eye, you may literally see stars. Compared to stimulation by light, this way of excitation needs much more energy, as the painful experience that comes with it will tell you.

Now, what are the adequate stimuli for our **chemical senses**? Humans have two sensory systems for chemicals, the tongue and the nose. The tongue is designed to respond to chemicals we encounter when making direct contact, while the nose is fine-tuned for chemicals that are volatile enough to be transported through the air. We possess six types of taste cells on our tongue: each specialized to detect one of six specific taste qualities–sweet, sour, salty, bitter, umami, and fat. Five of these cells respond to chemicals that are obvious, but what is umami? Umami cells responds best to **amino acids**, which are the building blocks of proteins. Twenty to twenty-five receptor cells are clustered together in taste buds. The **taste buds** are all over our tongue, but concentrated at the **papillae**. Each taste bud has a few cells of each type. The situation for our nose is similar, but more complex, because humans possess 300–400 different types of sensory cells for smells. The sensory cells are located in a small patch of tissue at the top of your nose, closest to the brain. Olfactory sensory cells look very much like typical neurons, consisting of the **cell body** where the nucleus is found, and **dendrite**, and a long **axon**. In olfactory receptors, the cell's dendrite sticks into the mucus of the nose so that it can detect incoming odorant molecules. The dendrite is the location where all the electrical events in the cell start when odor molecules are detected. On the opposite side of the cell is the long, thin axon extending all the way to the brain. Like in an electrical wire, any electrical signal generated by the sensory cell is transmitted to cells in the brain. While all olfactory neurons look the same under the microscope, each type has just one type of **receptor protein** on the surface of its dendrite. Each cell can respond only to the very few chemicals that interact with its receptor. The best picture we have for that interaction is a key that can start an engine. Each chemical has a certain shape. If that shape fits into a pocket of the receptor molecule, the transduction cascade is started. If a chemical does not fit any receptor protein, nothing happens. A small group of molecules that all have a similar shape will all lead to a response from the same receptor. This makes our sensory cells highly specific.

Let us now investigate what we know about **pheromones**, which are well known throughout the animal kingdom. For example, a female moth may release pheromones that the males use to find her. That works extremely well and some males can smell and find a single female that is several miles away. Mammals also use pheromones; for example, dogs, cats, beavers, and many more mark their territories with chemicals excreted from specialized glands or contained in urine. For the topic at hand it is important to know that for a multitude of mammals, we know at least a few of these pheromonal messages induce or influence sexual behavior. In fact, many animals signal their readiness for reproduction via such pheromones. Scientists also found that while mice use pheromone messages, they are not perceived within the normal sensory cells in the nose, but by two, separate, sac-like structures in the nasal cavity called **VNO** (vomeronasal organ).

Biologically, we humans share many similarities with other mammals, so there is good reason to ask, "Do humans also have a VNO? Is it fully functional?" In 50-90% of all adults (depending on the study), there is a small blind sac at the base of the nose, exactly in the location where one would expect the VNO. So the answer is most likely, yes, humans do have a VNO. Is it fully functional? If yes, then sensory cells in the VNO need to have dendrites with receptor proteins sensitive to chemicals on their surface so that the engine–key principle–can work. Additionally, the axons of the sensory cells need to go all the way to the brain, or no message can be transmitted. Usually in olfactory tissues, one can detect such bundles of axons. However, microscopic investigations of human VNO cells have revealed that only very few cells resemble chemoreceptive neurons. Moreover, no axon bundles have been found connecting the VNO with the brain. This makes it very unlikely that there is a role of the VNO in human pheromone communication.

Knowing that, can we now conclude the advertisement is plain nonsense? Well, let's be careful and look at all the data. As a college student, you may live in a dorm. If you are a female, then you might observe some interesting physiological events that comes along with living with other women. In 1971, Martha McClintock published a study that showed a large percentage of females living together in dorm rooms had synchronized menstrual cycles (1). She found that after about 5 months, about two-thirds of all female students living together begin their menstruation on the same day. Yet, at the beginning of the semester, there was no correlation at all. Nearly 30 years later (1998), Stern and McClintock demonstrated that this effect was caused by an odorless chemical secreted from cells in the armpits (2). Is there any evidence that females can be attracted to males by pheromones? Since 2000, two **steroids** were discovered that may have pheromone-like effects on the mood of females. These steroids occur in the armpits of males. They are sensed differently by men and women, and they may trigger a change in mood or prolong an already present positive feeling in females (3).

Therefore, there is good evidence that pheromone-like chemicals, indeed, can influence humans, even though they do not have a functional VNO. The advertisement still is completely overstated. Obviously, there still is a big gap between the scientific results and the claims of some companies selling pheromones. There is no single stimulus that alone can trigger any behavioral response; many important factors influence our feelings and behavior. For example, how attractive we rate a fellow human depends on how they look, the tone of their voice, body language, age, and character traits. If odors have an influence at all, then it is a minute part of a big picture. There is certainly no chemical out there that makes you fall in love with someone. So save your money, and buy your significant other flowers or his or her favorite DVD.

Questions about this case:

1. What is a pheromone?
2. What might be a hypothesis that could explain why humans might have phermones, even though we have no functional VNO.
3. The case says that humans have 300 to 400 sensory cells for smells, each with only one chemical receptor for odorant molecules in it's surface. Does that mean humans can only smell only 300 to 400 different smells?

Questions to go deeper:

1. What is the "adequate stimulus" for each sensory organ?
2. What are the advantages and disadvantages of communication by sight, sound, touch, and smell?

References:

McClintock, M. K. (1971). Menstrual synchrony and suppression. Nature 291, 244–245.

Stern, K., and McClintock, M. K. (1998). Regulation of ovulation by human pheromones. Nature 390, 177–179.

Jacob, S., and McClintock, M.K. (2000). Psychological state and mood effects of steroidal chemosignals in women and men. Hormones and Behavior 37, 57–78.

Additional reading:

http://www.hhmi.org/senses/

http://www.cf.ac.uk/biosi/staff/jacob/teaching/sensory/pherom.html

http://www.leffingwell.com/olfaction.htm

Attack of the "Killer" Tomato

A Case Study of Genetically Modified Foods

In 1994, Calgene Corporation introduced a revolutionary new food. The **Flavr Savr** tomato promised to be a "new generation" for food technology, and it was even approved by the **Food and Drug Administration (FDA)**. Traditionally, tomatoes bound for the supermarket had to be picked while they were still green so that they could maintain their firmness upon arriving at the market. These tomatoes were then sprayed with **ethylene** (their natural ripening hormone), so that they would have their "ripe" red color. This process produced a firm, red tomato for the customer; however, these tomatoes lacked the flavor of a fully vine-ripened tomato. Flavr Savr tomatoes promised to have the same firmness of the traditional "store" tomato, but the flavor of a fully vine-ripened tomato. How is this possible? If a tomato is picked ripe, shouldn't it rot by the time it reaches the store? What is going on? The answer is found in a new technology that allows scientists to create plants that have been genetically altered. Research scientists have discovered that they can both introduce new genes into plants and delete existing genes. By doing so, they can create plants that have been genetically modified, permanently. It turns out that the Flavr Savr is not a "regular" tomato since it is a product of this new technology. The Flavr Savr was the first **genetically modified (GM)** food introduced to the public. Scientists involved in creating GM foods have made some amazing claims. These claims range from promising new rot-resistant foods to providing drought-resistant plants. This new generation of food technology is living up to its hype, but it has encountered a flurry of public concern. Are these foods safe to eat? Are they toxic? Will they cause allergies? Will they harm the environment?

In order to address these concerns, we need to discuss how scientists alter these plants and the mechanism behind **gene** expression. The first question that you may be asking is what is a gene, and what does it do? It turns out that a gene is a unit of **DNA** that carries the genetic information to produce a single **protein**. Cells have a variety of tasks that need to be preformed so that they can survive and reproduce. In order to accomplish all of these tasks, the cell must make a variety of specialized **proteins**, all of which serve a specific function in the cell. However, if the cell is going to make these proteins correctly, it needs to have specific instructions for each one. These instructions are provided by each gene, and the entire **genome** serves as the instruction manual for the cell. In the cell, this instruction manual is located in the **nucleus**, in tightly packed **chromosomes**.

When the cell needs to produce a specific polypeptide, it will go through a two-step process. First it will form a **messenger RNA (mRNA)** copy of the gene through a process called **transcription**. The mRNA is simply a single-stranded copy of a gene that can be transported out of the nucleus and into the cytoplasm, where proteins are made. When this single-stranded mRNA is made, the double-stranded DNA has to be opened up, so that a copy can be created. Once the mRNA is exported to the **cytoplasm**, **ribosomes** will read the code in the mRNA and translate the message into a sequence of **amino acids**. This process is called **translation.** You can somewhat compare this process to building a house. When a house is being built, the architect first draws up blueprints of what needs to be built (the gene). The original plans are then copied (mRNA) and given to the construction team (ribosome), so that they can build the house (protein). The architect never gives the original plans to the construction team because they are expensive and hard to make. If the originals were damaged on the "job," they would be lost forever, and the house would not be built as it was designed to be. The same holds true for DNA. If the original gene were to be damaged during translation, a functional polypeptide could never be produced again. It is much "safer" to copy the gene first through transcription, and then to translate the copy.

So how does this apply to the Flavr Savr? Well it turns out that one gene is responsible for softening and ripening tomatoes: the gene that codes for the protein **polygalacturonase.** Normally during tomato formation, the plant makes proteins and transports a lot of carbohydrates to the fruit, so that it grows. Once the tomato has reached a certain size (usually at a certain time of the year), the plant will begin to make polygalacturonase protein in the tomato. This causes it to begin to ripen. Armed with this information, Calgene figured that if they could somehow make a tomato that produced much less polygalacturonase, it should take much longer to ripen. Thus, they could leave the tomato on the vine longer and it would still not be completely ripe by the time it reached the store. Now, they just needed to figure out how to do this. After extensive research, Calgene decided to try a new technique as a means of reducing the amount of polygalacturonase protein in the tomato. Basically, the technology works by inhibiting the translation of the polygalacturonase mRNA by the ribosome. Therefore, little or no protein is produced because the mRNA can't be translated.

In order to accomplish this, they first had to **clone**, or make an exact copy, of the polygalacturonase gene. Using tomato DNA, they cut out the polygalacturonase gene from the genome using **restriction enzymes**. Restriction enzymes are basically molecular scissors that cut DNA at specific sites. They then took this gene piece and placed it into a bacterial **plasmid** in the direction opposite to how it is normally found in the tomato genome. A plasmid is a small, circular piece of DNA that is normally found in bacteria that contains genes that are not on the bacteria's main chromosome. Once the polygalacturonase gene has been placed into the plasmid backward, it is then **introduced back** into ***E. coli***. By placing the plasmid into the bacteria, it allows it to serve as a host for the **recombinant DNA**. After the *E. coli* has acquired the recombinant plasmid, it is injected into a small tomato seedling. During this process, the *E. coli* transfers the plasmid to the plant cells. Thus, these modified plants not only have their normal set of chromosomes, but they also receive a bacterial plasmid that contains a reverse copy of the tomato polygalacturonase gene.

So how does this stop protein production? Well, it turns out that the plant is able to make mRNA from the plasmid, as well as from its chromosomes. Therefore, the plant can make mRNA of the backward polygalacturonase gene, as well as from the regular gene. This has some interesting results. When this "backward" mRNA is produced, it is the exact complement of the normal polygalacturonase mRNA. As a result, this "backward" **antisense** mRNA binds to the normal **sense** mRNA. This produces double-stranded RNA, which is similar to the normal "state" of DNA. Because ribosomes cannot translate the message from double-stranded RNA, protein is not produced. Therefore, very little, if any, polygalacturonase protein is produced.

Despite meeting its goal, Calgene pulled the Flavr Savr from the marketplace in 1997 because of massive public concern. Interestingly, there was no reported incidence of sickness or death from the tomato. Today, genetically modified foods are grown all over the world. Drought-resistant, insect-resistant, and insecticide-resistant soybeans, corn, cotton, and canola have all been produced and are being grown commercially all over the globe. In fact, in 2005, genetically modified crops were being grown by 8.5 million farmers in 21 countries. Ninety percent of these farmers are located in resource-poor developing countries. These crops allow poor communities to grow necessary food crops that would not normally survive otherwise in their environments. It has been estimated that 75% of all processed foods in the United States contain GM plant ingredients. Even with all of this consumption, there have been no reported illness, allergies, or deaths. There is some concern ecologically, however, because overuse of GM plants may lead to a lack of crop diversity. If farmers in a particular area are growing crops of only one variety, there could be a catastrophic effect if something goes wrong with that crop. A new environmental condition or insect pest could decimate the entire crop of the entire area.

In light of all the potential good GM crops could do for countries that need stable food crops, do you find it ironic that the first GM food introduced allowed people in the United States to eat a better tomato?

Questions about this case:

1. What are some of the benefits of GM foods? don't rot & good for poor countries
2. Are there any downfalls to GM foods? can be destroyed
3. Should food products that contain GM plant ingredients be labeled as such?

Questions to go deeper:

1. How does this technology relate to gene therapy? adding & subtracting genes
2. What are some other genetically altered foods, and why are they being created? cotton soybeans
3. How have scientists altered cotton and corn to make them insect resistant? took gene from tabaco which is a natural insecticide

References:

http://dragon.zoo.utoronto.ca/~jlm2000/T0501D/introduction.html

The Linebacker Who Had No Hands

A Case Study of Development and Thalidomide Exposure

As the son of a football coach, I grew up watching a lot of football. Most of these games I watched from the sidelines, and needless to say, I saw many weird and unusual things happen. But, nothing really compared to a college game I watched one Saturday afternoon when I was in high school. I distinctly remember watching the quarterback throw a pass over the middle, only to have a linebacker tip the ball up in the air and then proceed to intercept it. The shocking part was when the linebacker jumped up to celebrate and threw his arms in the air, he had no hands! I watched the guy the rest of the game and found that he was incredibly talented. Why would this surprise me? I just had never seen anyone play football without hands. I found myself wondering how he had lost his hands.

I only later found out that this talented linebacker did not lose his hands; he was born without them. As I began to think about how this may have happened, I remembered a horrible tragedy that occurred in the 1950s and 1960s because of a drug called **thalidomide**.

Thalidomide was first discovered and synthesized in West Germany in 1953 by the pharmaceutical company Chemie Grünenthal. After multiple animal tests, the drug was found to be a semipowerful sedative. Clearing all of the animal tests preformed, thalidomide was cleared for sale in Europe in 1957, and was immediately placed on the market. Marketed as a sleep aid and a cure for morning sickness for pregnant women, thalidomide was a hit. However, a little time passed and reports began to come in associating thalidomide with birth defects. Even with the reports, thalidomide continued to be sold in Europe and throughout the world. However, in 1961, thalidomide was found to have **teratogenic** effects on human development, especially if it was taken during the first 25 to 50 days of pregnancy. Soon thereafter, it was removed from the shelves. Sadly, nearly 15,000 children were born with thalidomide-related birth defects during this time period in over 40 countries. In order to understand how thalidomide causes birth defects, we must look at human development.

Human development is actually a step-wise process. It begins with fertilization and ends with birth. This step-wise process is usually broken down into stages, which are determined based upon **morphological** changes that occur in the embryo, not age or time. Basically, development begins at fertilization, and continues as the single **zygote** begins to divide and grow. This division leads to a **blastocyst** with two distinct layers: the outer layer, called the **trophoblast**, and an inner compact ball of cells called the **inner cell mass**. The trophoblast is composed of cells that will eventually help to give rise to the **placenta**. The cells of the inner cell mass will eventually give rise to the **fetus**. Each cell in the inner cell mass is called an **embryonic stem cell**. These cells are important because they are not yet **fated** to become a specific type of tissue or organ. Each embryonic stem cell can still become any type of cell, tissue, or organ that is found in the body. Because of this potential, scientists are currently trying to learn more about stem cells as a potential cure for many diseases. However, because human embryos have to be destroyed to obtain embryonic stem cells, many groups are morally opposed to this type of research, especially since stem cells can be obtained from other sources (i.e., the placenta and **adult stem cells)**.

After the formation of the blastocyst, the embryo then goes through a process called **gastrulation**. Gastrulation is the process by which the cells of the ball-like embryo change their location in order to give rise to the three **germ layers** of the body. The first germ layer, the **ectoderm**, will eventually give rise to the skin, hair, and the nervous system. The cells of the **mesoderm** are fated to give rise to the bones and muscles, and the **endoderm** gives rise to organs such as the stomach, intestines, and the lungs (among other things). Once the germ layers have been established, cells in the embryo continue to divide and grow, and the cells in the three germ layers begin to form the structures described above. All of this occurs during the first 18–19 days of development. During days 19–21 of development, the nervous system begins to form through a process called **neurulation**. During this time, the heart is also forming, and begins to beat at around day 22. Other developmental highlights include the formation of all major organs, including the eyes and the limbs during days 23–53. After day 53, the **fetus** does not undergo major changes to its morphology; it just grows and continues to develop until birth. So basically, by the end of the 8th week (two-thirds of the way through the **first trimester)** you have a fetus that has formed all of its tissues and organs. The rest of the first trimester and the **second** and **third trimesters** of pregnancy are really just periods of growth.

Now back to thalidomide. Thalidomide exposure during development usually leads to two types of birth defects: **amelia** and **phocomelia**. In both cases, limb development is affected. Children born with amelia completely lack one or more of their limbs, while people born with phocomelia are born with short, flipper-like limbs.

Thalidomide causes these types of birth defects because it inhibits **signaling** in the developing **limb bud.** In vertebrates, limbs form out of a nodule that is formed on the body early in development. Cells in these nodules first secrete different molecular signals to orient the limb buds to the rest of the body. Once this occurs, another signal is secreted that causes the cells in these areas to divide, so that the limbs can grow. Soon after this point, the cells at the tip of the limb buds become fated, so that they will give rise to the hands. When the limb buds are exposed to thalidomide, it blocks the signal that causes limb growth, but it has no effect on the signals that lead to cell fate. Therefore, when the embryo is exposed to thalidomide, the cells in the limb buds do not divide, so the limbs are not able to elongate. As a consequence, children exposed to thalidomide have short (or no) limbs, depending on when they are exposed to it (between days 23–53 of development). The earlier the exposure, the more likely that they would be born without any limb at all because the limb bud is never really allowed to be formed. Sadly, this problem was not seen until it was too late.

Wait a minute, if thalidomide was tested on animals, why was this problem not seen during testing (especially if vertebrates develop limbs in much the same way as we do)? It turns out that Chemie Grünenthal did not do the proper testing. Even though thalidomide was marketed as a pregnancy drug, it was not originally tested on pregnant animals. These tests were not done until the early 1960s, when Chemie Grünenthal petitioned to have the drug sold in the United States. Because of the negative reports around the world, the **Food and Drug Administration (FDA)** postponed approval for U.S. sales until further testing could be done. Had the tests been done earlier, the teratogenic effects of thalidomide would have been discovered before the drug ever hit the market in other areas. Since this tragedy, much more stringent testing requirements must be met by every drug that is intended for human consumption. Hopefully, these new requirements will prevent a tragedy such as this from ever happening again.

So what about our amazing linebacker? You may now be thinking how his symptoms could have been caused by thalidomide? If he would have been exposed to thalidomide, either he would have lacked arms or he would have had flippers. Also unless he was born somewhere other than the United States, he could have never come into contact with thalidomide. The linebacker was only missing his hands. Therefore, his birth defect must have been caused by something completely different, but that is the topic of a whole other case.

Questions about this case:

1. What are other drugs that have been found to be teratogenic?
2. What are other organs that are formed by the endoderm?
3. What are adult stem cells, and are they as useful as embryonic stem cells?
4. What is the typical gestation period for humans, and how is that divided up into trimesters?

Questions to go deeper:

1. What is a teratoma? What does it look like?
2. What is the current FDA procedure for approving a new drug?
3. What are some possible causes of our linebacker's birth defect?

References:

http://www.wikipedia.org/wiki/Thalidomide

http://www.embryology.med.unsw.edu.au/embryo.htm

How Large Can a Brain Tumor Get?

A Case Study of the Nervous System

Deb did not have any obvious symptoms until her seizure. It happened quickly; she did not even realize what was happening. With her husband at work, she was at the park with her three kids. She simply passed out and had what would appear to be a minor seizure. But, whether the seizure was major or minor, people do not have seizures for no reason. Deb was rushed to the hospital as friends were called to care for her kids. The emergency room at the local hospital was full, so Deb was taken to a small clinic. I went by the clinic later that evening to visit with Deb's husband, Todd. He said the doctors had done an MRI and they had found something on the scan but were not certain what it was. Deb was transferred to a hospital where an image was generated by **MRI**. Ultimately, Deb was diagnosed with a stage 2 astrocytoma; a cancer in her brain originating in a cell called an **astrocyte**.

When you look at the scan of Deb's brain, you realize that it does not take a physician to see the tumor. What astounds everyone who sees the scan is how big the cancer got with almost no symptoms. As Deb and Todd thought back over the previous couple of months, they began to realize that there really were symptoms, but they formed so gradually that no one was really concerned. As the cancer had formed in, and pushed on, the part of Deb's brain that was involved in speech, both Todd and Deb's friends could tell that she was having an increasingly hard time finding the correct words to talk about very simple things. This particular tumor was slow-growing, but you could imagine that if it had formed in other parts in her brain, the symptoms might have shown up much earlier. To understand how the symptoms for Deb's tumor might have been different if it had grown in a different location, we need to take a look at the functions of the different portions of the brain.

The **brain** is part of the section of the nervous system known as the **central nervous system (CNS)**, which also contains the **spinal cord**. This system is different from the **peripheral nervous system (PNS)**, which includes all the **nerves** in the nervous system that run from the brain and spinal cord out to the various **sensors** that receive the inputs from the environment, as well as the **effectors,** which the brain controls. The brain is composed of five major parts, each of which has specific functions that it controls. However, it must be remembered that no part of the brain really controls anything on its own. Rather, certain parts of the brain are in charge of certain events, but other parts play secondary or tertiary roles, also.

At the base of the brain, where the spinal cord enters into the brain itself, is the **medulla oblongata**. This section of the brain, frequently called the **brain stem**, is in control of many of the most basic bodily functions. It is in charge of making sure our heart beats correctly, our lungs inhale and exhale, our blood pressure is kept stable, and such things as swallowing occur. It also controls such nuisanced things as vomiting, coughing, sneezing, and hiccupping. While this latter group may be things we can do without, the former groups are essential for life. If Deb's tumor had been in this area of the brain, her life would have been greatly impacted, and her life may actually have been significantly shortened due to the importance of this brain portion and what it controls.

Just to the rear of the medulla is the two-lobed section called the **cerebellum**. If Deb had a tumor here, she would have started having all sorts of muscle problems, because this is the section of the brain that has primary control of our skeletal muscles and the things they relate to. For example, while this includes muscle tone, it also includes our posture, balance, and coordination. If Deb had had troubles learning to catch a ball, for example, this would have been the area being impacted by the tumor. Again, since something like coordination requires coordinating input from a variety of sensors as well as the muscles, it requires several parts of the brain for us to do something like catching a ball or throwing a pass.

In the center of the brain are several important sections that would be affected by a tumor growing inside the brain itself. In the middle of the brain is the **thalamus**, a small but important section of the brain. Since many of the **neurons** coming into the brain run through this section, it serves as a relay station, sending the various nerve impulses coming into the brain to the correct sections of the brain. If you have ever seen the old-fashioned telephone switchboards with all the various wires that are used to send the phone call to the proper telephone, you get an idea of how this section works. However, it also serves in helping with our memory and controls some of our emotions.

Just below the thalamus is the **hypothalamus**, which is in control of another very important aspect of our body called **homeostasis**. This is the process that makes sure that all the important things in our body are kept at their proper level or in balance. To do this requires controlling many of our body systems, as well as telling us when we are hungry, thirsty, or sleepy. It helps maintain our internal body temperature and our water balance, which affects our blood pressure. This portion of the brain also controls the **pituitary gland**, located at the base of the brain just behind our eyes, which is called the master gland because it controls so many other glands in our body. Many of the emotions and aspects related to reproduction are also controlled through this brain section. The presence of a tumor in this section of the brain, or in the thalamus, would probably have been recognized relatively quickly, simply due to the location of the sections. Whereas a tumor affecting the other three brain parts could be located on the outside of the brain, allowing some give on the part of the brain as the tumor grew, a tumor located in the center of the brain would not have that advantage.

The largest portion of the brain, what makes us biologically different from all other animals, is the **cerebrum**. It is this section of the brain that was being impacted by the tumor that Deb was suffering from. It is divided into two easily seen **hemispheres**, the right and left, and into two layers, the **cortex**, the outer layer, and the **white matter**, the inner section of the brain. The cerebrum is also broken into four different **lobes**, each of which has primary control over certain functions. At the very front of the cerebrum, right behind the forehead, are the **frontal lobes**. This is the specific area of the brain that was being affected by Deb's tumor, as it is the primary area of the cerebrum controlling speech and short-term memory. It also has to do with initiating the voluntary commands that we send out to various parts of our body and works with the cerebellum to coordinate much of our body's movements. Complex motor skills that have to be learned, such as playing an instrument or writing, are controlled from here, as is thought, concentration, and our abilities to solve problems and make future plans.

Immediately behind the frontal lobes are the **parietal lobes**, which receive input from the various sensors located in the skin and in our joints. This helps people to know where they are, the position of their body parts, and to be able to gather information on the shapes and texture of objects. It also functions in our sensation of taste and in our spatial relationships with things around us. It thus works with the adjacent **occipital lobes**, which control virtually everything related to vision, to help us with our visual perceptions. As with all the sections of the brain that control various senses, the occipital lobes not only work in allowing us to see, but also to interpret, remember, and relate those images to other senses.

Finally, the **temporal lobes** control the auditory, or hearing, sense. They also control this sense's relation to our ability to respond to sounds and images so that we can relate all these different things. Finally, these lobes are important to our emotions and memory, allowing us to take a current event and stick it into our short- and long-term memory banks.

While Deb's tumor made its presence known in a rather dramatic way, we become aware of many tumors due to a slowly accumulating series of problems. Once a doctor is made aware of these problems, they can frequently predict which area of the brain is being affected from the symptoms being demonstrated. This is also a useful technique when people suffer from **strokes**, which occur when a blood vessel in the brain becomes blocked for some reason. As the brain is deprived of oxygen, it begins to die, and whatever that portion of the brain controls is therefore impacted. Doctors can tell approximately where in the brain the stroke occurred by testing to see exactly what brain functions are impaired.

The idea of determining where tissue damage has occurred by looking at the symptoms holds true for the spinal cord, as well. If a catastrophic injury to **vertebrae** in the spinal column occurs where the spinal cord is damaged, any nerve impulses that would usually travel through that point in the spinal cord will cease. So if the spinal cord is severed by an injury, we can predict with some certainty where this has occurred by where the patient can feel and what muscles can be moved. To make this type of diagnosis easier, doctors have divided the 33 human vertebrae into 4 groups and named them according to their location in the body. Starting at the base of the skull there are 7 **cervical vertebrae**, 12 **thoracic vertebrae**, 5 **lumbar vertebrae**, 5 that are fused together to make the **sacrum**, and 4 that make up the **coccyx** (tail bone). An injury that severs the spinal cord at C1 through C3 would result in the loss of the ability to control the **diaphragm** and would result in either death or life permanently on a respirator. The spinal cord from C5 through T1 sends out the peripheral nerves that control the arms. A break at C5 would result in the person being a **quadriplegic**, which means they would have no use of arms or legs. But, spinal cord injuries from C6 through T1 would result in partial loss of arm movement and sensation, depending on what nerves were affected. The legs are controlled by nerves that exit the vertebral column at L1 through L5, so a break in the spinal cord at L1 would leave a person without movement of his or her legs, which is commonly referred to as being a **paraplegic**. Clearly, a person can break a vertebra but not damage the spinal cord, but when these injuries occur, great care must be taken to stabilize the person so that merely moving the person doesn't cause spinal cord damage.

The situation with Deb was complicated. The best intervention for Deb's cancer was surgery, but the operation was tricky. The physicians wanted to get as much tumor out as possible, without harming Deb's ability to speak. Ultimately, due to its location and involvement with the brain, not all of it could be removed. She is currently back at home with her family and doing very well. She is undergoing rehabilitation to recover abilities lost to the damage done by the tumor and she gets regular MRIs to monitor how the cancer is growing. Hopefully, over time, additional surgeries or treatments will help Deb win this battle.

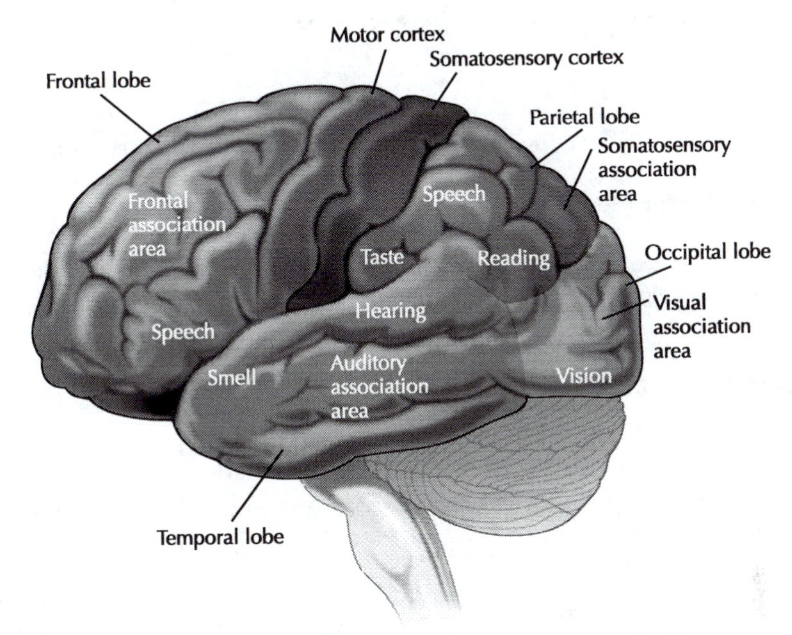

This illustration shows the lobes and major functional areas in the cerebrum.

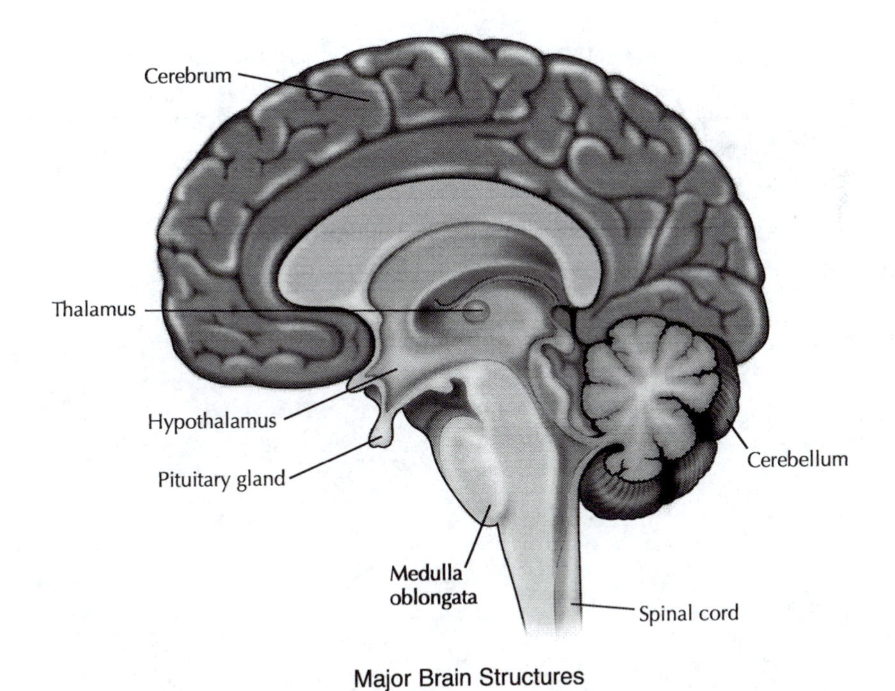

Major Brain Structures

Questions about this case:

1. Sometimes head injuries cause bleeding and swelling in the brain. Even though the brain itself may not be damaged, this can be quite dangerous. Explain why.
2. What is the blood–brain barrier?
3. Often when a person has a minor to moderate stroke, you see only one side of the body affected, why is that?

Questions to go deeper:

1. The case indicates that sensory stimulus coming in from a peripheral nerve is sent up the spinal cord to the brain. How would this differ from a stimulus that causes a reflex?
2. What are vertebral disks, what do they do, and what happens when they "rupture"?
3. The brain is covered with a set of membranes called meninges. How are these related to meningitis?
4. What does the pituitary produce that controls so many functions in the body?
5. There are a number of diseases that involve the selective degeneration of neurons in the spinal cord. A good example is amyotropic lateral sclerosis (ALS). What cells degenerate in ALS, and what are the symptoms?

Reference:

http://www.merck.com/mmhe/sec06/ch076/ch076b.html

Index

A

B

C

D

E

F

G

H

I

J

K

L

M

N

O

P

Q

R

S

T

U

V

W

X

Y

Z